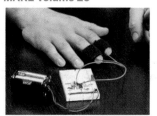

Make: Volume 27

ROBOTS!

Family fun project

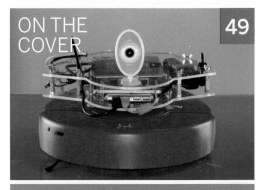

THE SPY WHO VACUUMED: Hacked with a router and webcam, this Roomba spies via the internet. Creative direction by Sean DallasKidd. Photography by Garry McLeod. Prop styling by Tietjen Fischer.

SPAZZI: A bouncy, solenoid-powered robotic character from the makers of Keepon.

TOY HACKS: DJ Sures shows you how to make interactive robots out of common toys.

Vol. 27, July 2011. MAKE (ISSN 1556-2336) is published quarterly by O'Reilly Media, Inc. in the months of January, April, July, and October. O'Reilly Media is located at 1005 Gravenstein Hwy. North, Sebastopol, CA 95472, (707) 827-7000. SUBSCRIPTIONS: Send all subscription requests to MAKE, P.O. Box 17046, North Hollywood, CA 91615-9588 or subscribe online at makezine.com/offer or via phone at (866) 289-8847 (U.S. and Canada); all other countries call (818) 487-2037. Subscriptions are available for $34.95 for 1 year (4 quarterly issues) in the United States; in Canada: $39.95 USD; all other countries: $49.95 USD. Periodicals Postage Paid at Sebastopol, CA, and at additional mailing offices. POSTMASTER: Send address changes to MAKE, P.O. Box 17046, North Hollywood, CA 91615-9588. Canada Post Publications Mail Agreement Number 41129568. CANADA POSTMASTER: Send address changes to: O'Reilly Media, PO Box 456, Niagara Falls, ON L2E 6V2

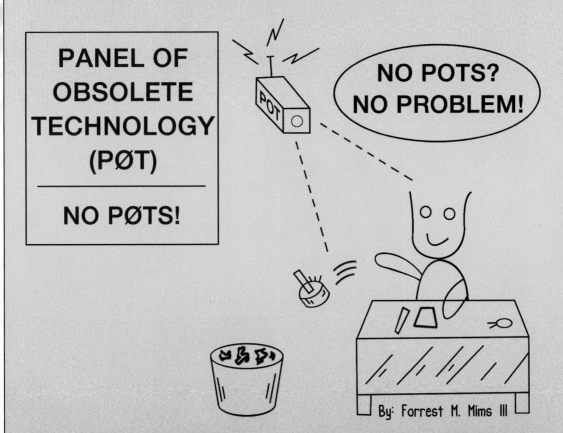

Make: Projects

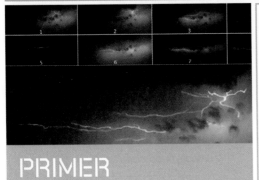

Make: Volume 27

Maker

MOVIE MAKER: *Avatar* F/X innovator Glenn Derry shares his recipe for an indie-budget virtual camera.

DIY

HOT WATER: How will you provide enough safe drinking water for your family during a nuclear crisis?

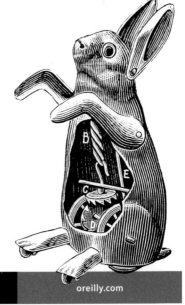

> "When I'm working on a problem, I never think about beauty. I think only how to solve the problem. But when I have finished, if the solution is not beautiful, I know it is wrong."
> —Buckminster Fuller

technology on your time

FOUNDER, GM & PUBLISHER, MAKER MEDIA
Dale Dougherty
dale@oreilly.com

EDITORIAL

EDITOR-IN-CHIEF
Mark Frauenfelder
markf@oreilly.com

EXECUTIVE EDITOR
Paul Spinrad
pspinrad@makezine.com

MANAGING EDITOR
Keith Hammond
khammond@oreilly.com

SENIOR EDITOR
Goli Mohammadi
goli@oreilly.com

STAFF EDITOR
Arwen O'Reilly Griffith

EDITOR AT LARGE
David Pescovitz

DESIGNER
Katie Wilson
kwilson@oreilly.com

PRODUCTION DESIGNER
Gerry Arrington

PHOTO EDITOR
Sam Murphy

COPY EDITOR
Gretchen Bay

ASSISTANT EDITOR
Laura Cochrane

ADMINISTRATIVE COORDINATOR
Ramona Minero

PUBLISHING

MAKER-IN-CHIEF
Sherry Huss
707-827-7074
sherry@oreilly.com

SENIOR SALES MANAGER
Katie Dougherty Kunde
katie@oreilly.com

SALES ASSOCIATE
PROJECT MANAGER
Sheena Stevens
sheena@oreilly.com

SALES & MARKETING
COORDINATOR
Brigitte Kunde
brigitte@oreilly.com

ASSOCIATE PUBLISHER & GM,
E-COMMERCE
Dan Woods
dan@oreilly.com

DIRECTOR, RETAIL MARKETING
& OPERATIONS
Heather Harmon Cochran

OPERATIONS MANAGER
Rob Bullington

DIRECTOR, PRODUCT
DEVELOPMENT
Marc de Vinck

MAKER SHED EVANGELIST
Michael Castor

ONLINE

DIRECTOR OF DIGITAL MEDIA
Shawn Connally
shawn@oreilly.com

EDITOR-IN-CHIEF
Gareth Branwyn
gareth@makezine.com

ASSOCIATE EDITOR
Becky Stern
becky@oreilly.com

DIRECTOR OF TECHNOLOGY
Stefan Antonowicz

WEB PRODUCER
Jake Spurlock

EDITOR AT LARGE
Phillip Torrone
pt@makezine.com

COMMUNITY MANAGER
John Baichtal

PUBLISHED BY

O'REILLY MEDIA, INC.
Tim O'Reilly, CEO
Laura Baldwin, President

Copyright © 2011
O'Reilly Media, Inc.
All rights reserved.
Reproduction without
permission is prohibited.
Printed in the USA by
Schumann Printers, Inc.

Visit us online:
makezine.com

Comments may be sent to:
editor@makezine.com

CUSTOMER SERVICE

cs@readerservices.
makezine.com

Manage your account online,
including change of address:
makezine.com/account
866-289-8847 toll-free
in U.S. and Canada
818-487-2037,
5 a.m.–5 p.m., PST

Follow us on Twitter:
@make
@craft
@makerfaire
@makeprojects
@makershed
On Facebook: makemagazine

MAKE TECHNICAL ADVISORY BOARD
Kipp Bradford, Evil Mad Scientist Laboratories,
Limor Fried, Joe Grand, Saul Griffith, William Gurstelle,
Bunnie Huang, Tom Igoe, Mister Jalopy, Steve Lodefink,
Erica Sadun, Marc de Vinck

CONTRIBUTING EDITORS
William Gurstelle, Mister Jalopy, Brian Jepson, Charles Platt

CONTRIBUTING WRITERS
William Abernathy, Alex Andon, Doug Bradbury,
Raymond Caruso, Abe Connally, Kindy Connally-Stewart,
Cory Doctorow, Adam Flaherty, Frank Ford, Bob Goldstein,
Saul Griffith, Rachel Hobson, Andrew Lewis, Frits Lyneborg,
Gordon McComb, Immanuel McKenty, Marek Michalowski,
Forrest M. Mims III, Philip "Corky" Mork, Pauric O'Callaghan,
Bob Parks, Riley Porter, Stacey Ransom, Jerry Reilly,
Chris Singleton, Bruce Stewart, Jerry James Stone,
Doug Stowe, DJ Sures, Peter Tabur, Cy Tymony,
Tom Vanderbilt, Rose White, Adam Zeloof,
Robert M. Zigmund, Lee David Zlotoff, Zach Zundel

CONTRIBUTING ARTISTS
Roy Doty, Nick Dragotta, Timmy Kucynda, Juan Leguizamon,
Tim Lillis, Brian McLaughlin, Garry McLeod, Rob Nance,
James Provost, Kathryn Rathke, Damien Scogin, Jen Siska,
Colin Way, Noah Webb

ONLINE CONTRIBUTORS
John Baichtal, Chris Connors, Collin Cunningham,
Adam Flaherty, Matt Mets, John Edgar Park,
Sean Michael Ragan, Matt Richardson, Marc de Vinck

INTERNS
Cameron Caywood (video), Eric Chu (engr.),
Craig Couden (edit.), Sydney Francis (mktg.),
Gregory Hayes (photo), Brian Melani (engr.),
Tyler Moskowite (engr.), Lindsey North (edu.),
Nada Raoof (online), Nick Raymond (engr.)

PLEASE NOTE: Technology, the laws, and limitations
imposed by manufacturers and content owners are
constantly changing. Thus, some of the projects described
may not work, may be inconsistent with current laws or
user agreements, or may damage or adversely affect some
equipment.

Your safety is your own responsibility, including proper
use of equipment and safety gear, and determining whether
you have adequate skill and experience. Power tools,
electricity, and other resources used for these projects
are dangerous, unless used properly and with adequate
precautions, including safety gear. Some illustrative photos
do not depict safety precautions or equipment, in order
to show the project steps more clearly. These projects are
not intended for use by children.

Use of the instructions and suggestions in MAKE is at
your own risk. O'Reilly Media, Inc., disclaims all responsibility
for any resulting damage, injury, or expense. It is your
responsibility to make sure that your activities comply
with applicable laws, including copyright.

CONTRIBUTORS

Alex Andon's (*Jellyfish Tank*) fascination with marine animals was sparked on a sail through the British Virgin Islands while free-diving underwater tunnels in coral reefs. He went on to work for the Sea Turtle Protection Society of Greece, camping on beaches in Crete to protect nests and hatchlings. Alex accumulated extensive experience in building aquariums for his own research projects at Duke and the University of Delaware. He founded Jellyfish Art on a platform of cutting-edge jellyfish husbandry techniques that have been developed over the past decade. Alex has a B.S. from Duke in biology and environmental science.

Marek Michalowski (*Spazzi: A Solenoid-Powered Dancebot*) has been excited by robots and airplanes since childhood, and things haven't changed much. Together with Hideki Kozima, designer of the robot Keepon, Michalowski co-founded BeatBots LLC in 2007. He's enthusiastic about the growth of desktop fabrication and community hackerspaces, and about their implications for the future of toys, art, and design. In his spare time, he enjoys using his private pilot, scuba, and motorcycle licenses (though never at the same time). Michalowski was born in Poland, grew up in New York, and currently lives in San Francisco.

As a kid, **Peter Tabur** (*Limelight*) had just about every construction set there was, and loved taking broken things apart to figure out how they worked. His love of making, fixing, and building is unabated, so it should come as no surprise that he went to engineering school. He lives in Augusta, Mich., with his wife, Teri, and foundling beagle, George. Peter likes to make pizza and furniture and loves his 1950s vintage Rockwell radial arm saw — "a massive cast-iron beauty" he fully restored. He's currently between jobs and is taking advantage of his ample free time to work on his long list of projects.

Gregory Hayes (MAKE photo intern) is a California photographer, writer, puppeteer, and handyman, among other things; it's his hope that this rare combination of skills will keep him high on the invite list for off-Earth colonization efforts. In the meantime, he enjoys canoeing, tinkering, cooking, and eating. He lives on the Russian River, throws sticks for his two wonderful dogs ("this never, ever gets old"), likes making tools on the fly ("particularly if I can reuse parts of other projects for new purposes"), and really, really loves artichokes.

Kathryn Rathke (MAKE author portraits) is "a practical Wisconsin girl" living in Seattle with her husband and her "gargoyle-faced dog," Bunny. Her portraits have appeared on the covers of magazines, been projected onto screens, turned into embroidery, and scaled down to fit onto cupcakes. She "treasures wit and absurdity," took up competitive badminton a year ago, and just had her first Friend Request from one of her subjects. She's learning Flash animation so she can teach her line portraits to blink. kathrynrathke.com

If **Colin Way** (*Teaching Old Toys New Tricks* photography) were to describe what he strives for in images, it might sound like this: illustrative and immediate, layered and detailed. Colin has been fortunate to work with a varied client list, including *Chatelaine*, Dell, WestJet, *Canadian Business Magazine,* and *Fashion*. One of the best things about his job is the interesting people he meets along the way (including meeting DJ Sures for this issue). To see more of his work, visit colinway.com.

Building Robots That Play

THIS MAY, MAKER FAIRE BAY AREA brought together 100,000 people for a weekend to celebrate creativity, tinkering, and the DIY spirit. I'm proud to see our event grow so large in just six years. It's a huge indicator that the maker movement continues to broaden by inviting more people to see themselves as makers. Maker Faire is a tremendous outpouring of creative energy: it comes from diverse sources, including very personal motivations, and it goes in all directions.

A reporter at Maker Faire asked me to point out "useful inventions" and "practical applications." I had to think for a moment, but I replied that what dominates Maker Faire are things that are less practical and more playful. It's like seeing the world through the eyes of a kid, where ordinary stuff like cardboard has all kinds of possibilities for playtime. The power lies in your imagination.

We tend to look at invention and innovation as serious business, causing us to miss the bigger picture. In 1876, a committee evaluating Alexander Graham Bell's patent for the telephone dismissed the device as "impractical," "idiotic," "hardly more than a toy." (My favorite reason was that there was no shortage of messenger boys.) We have a hard time predicting how technology creates new opportunities for people to enjoy life.

Robots were conceived as laborers, designed to do the boring, repetitive tasks that humans didn't want to do. In a dystopian vision of the future, the machines rule because they do all the work and there's really no need for humans.

But can a machine be playful? Can it entertain us, even mimic us? Indeed, most of the technology in our lives we use for play, and that play personalizes the technology, whether it's cars or computers. I wonder how long it was before IBM recognized the popularity of solitaire on PCs.

Nothing could be more mundane than a robot that cleans house, as iRobot's Roomba does. But ever since the debut of the robotic vacuum cleaner, hackers have treated it like a plaything. This issue's "Roomba Recon" project (page 49) starts with a Roomba as a mobile platform, then adds wireless networking and a video camera to create a spy robot with all kinds of potential for mischief. Useful? Maybe. Fun? Definitely.

Our "Spazzi" bot (page 56) is a rhythmic response to the question, "So you think your robot can dance?" Developed by Marek Michalowski of BeatBots, Spazzi amuses us. Yet the thinking behind this toy robot reflects a serious application. BeatBots also developed the Keepon robot, a yellow ball that can produce a range of simple expressions for nonverbal interactions with children. Keepon has been used in studies of autistic children to understand how they react to social cues.

"Teaching Old Toys New Tricks" by DJ Sures (page 66) shows you how to use the Bluetooth EZ-B Robot Controller to transform a dopey Digger the Dog pull-toy into an autonomous robot that can chase a red ball and obey your voice commands. Someone will no doubt program Digger to chase cars as well.

The playful robots we feature in MAKE say a lot about us humans and how we create things in our own image. We've domesticated dogs to do things like play fetch, come when they're called, and lie by our side while we sit in our favorite chair. Like dogs, robots are becoming our companions, demonstrating the ability to learn new routines that make us happy. ◪

Dale Dougherty is the founder and publisher of MAKE.

Jacobean Joinery, Zinc Fumes, Space Pods, Stealth Networks

✉ Finally subscribed. (Me cheap, and the cover price put me off for a while.) This community, this attitude has given me a second technical life. I am a controls engineer looking for something after the current career ends.

Your attitude has been a cool spring breeze. It makes me love the basics of my science-fair youth and believe in open-handed sharing for the good of our common future.

I love the hell out of this MAKE thing. I'm on board.

—Dave Weidling, Arcata, Calif.

📷 The pattern for the "Fool's Stool" (Volume 26) repeats a common mistake: cutting notches in the aprons (the horizontal pieces). This creates a high likelihood of short-grain failure (splitting along the grain).

Instead, cut a slot only in the legs, and make it deep enough to accept the apron's full height.

Next, rather than gluing the joint, pin it. Assemble the 4 base pieces and hold them together with tape, clamps, or a helper. Now drill two ¼" holes through each joint, one high and one low, going from the outside edge of the leg through the apron and at least ½" beyond. Pound in a ¼" oak dowel and cut it off flush (the way the author affixes the seat).

The result will be a rock-solid frame you can expect to hold up a lifetime, if not 500 years! This is how all the surviving boarded stools from the 16th and 17th centuries that I've examined were made.

It's not only more authentic, it's faster, easier (4 slots to cut, not 8, and they needn't fit tightly because they won't be glued), and much stronger. See also: albionworks.com/Stools/STOOLS.htm.

—Tim Bray, Albion, Calif.

✉ In "Weekend Warrior Gravity Racer" (Volume 26), author Jeremy Ashinghurst says to "sand off the galvanized surface layer" from EMT tubing before welding, "as it can weaken the joint." The reason it weakens the joint is because burning zinc gives off hazardous gases. The simple answer is: don't weld galvanized steel. There are sources for thin-walled steel tubing without using EMT conduit.

I use a TIG welder, and occasionally I can't avoid welding on galvanized steel. Even after thorough grinding, the weld is never as good as on plain steel. In addition, the torch nozzle and tungsten get contaminated and spoil your next weld on steel or stainless steel.

Otherwise it's a great story — my compliments to Jeremy on an excellent cart design.

—Gord Martin, Mississauga, Ontario

EDITOR'S REPLY: Good tip, Gord. MAKE's Technical Advisory Board agrees: welding galvanized metal creates fumes that aren't good for you or your weld. Further, EMT conduit is dip-galvanized, so the zinc coating is also inside, where it's impractical to grind off.

📷 I teach astronomy, physics, and engineering, and I was inspired by one of my students to think about making a device he could use to look at the stars and planets. I hope your readers find it an interesting challenge.

This student is good with a joystick, as he uses one on his wheelchair. I envision an egg-shaped pod (remember the movie 2001?). He could sit in a comfortable (heated?) chair inside the pod away from the cool night air and wind. We have a 40lb, 20x power, 120mm Nikon binocular telescope that could be installed at eye level. The pod would be mounted on a motor-driven gimbal or wheels, controlled by the joystick, to point the student, pod, and telescope at any astronomical object of interest.

I'm experimenting with a microcontroller to begin the project, and I thought about a fiberglass pod, but I have no experience with fiberglass or carbon fiber.

—*Joe Polen, Redding, Calif.*

EDITOR'S REPLY: Joe, what a nice idea. Makers, see our Make: Arduino page (makezine.com/arduino) and carbon fiber primer (makeprojects.com/Project/c/880), and/or basic fiberglassing instructions (Google it). Then show us your Astronomy Pod at makeprojects.com!

Do you sell back issues? My buddy loves pinball machines, and I'd give him my copy of Volume 08, but I want to make sure I can replace it. Collecting MAKE is just as important (if not more so) than collecting comics.

Also, could MAKE do an issue on stilts? A cool company called Weta out of New Zealand makes "reverse leg" stilts for movies, and I've seen people make their own on YouTube, but there's little information on building them. Extreme stilts and/or hydraulic or pneumatic stilts would be awesome on so many levels.

Thanks for MAKE. It has changed my life, and may well speed up that singularity event author/scientist Ray Kurzweil writes about.

—*Robert Bohan, Kent, Wash.*

EDITOR'S REPLY: Robert, thanks for passing MAKE along. Yes, back issues are for sale at makershed.com; Volume 08 is out of print, but you can get it as a PDF. Better yet, as a subscriber, you can read, search, and download back issues online in the MAKE Digital Edition.

I love this magazine! Someone should start a charity to distribute MAKE to young, underprivileged nerds and geeks living in remote areas far from hackerspaces (like me, except I already get MAKE).

I really liked the conductive play-dough in Volume 22 as a way to safely teach children about circuitry. Also I loved "Maker: Kid

Robot" (Volume 23) and the recent "Arduino Revolution" issue (Volume 25). It's refreshing, encouraging, and inspirational to read about folks who are doing the stuff I wish I could do — and hopefully will do (I'm 17).

For future Maker Faires, I recommend Portland, Ore.; Boise, Idaho; and Spokane, Wash. Bring MAKE to the Northwest — you will have at least one happy maker!

—*Lizbeth, Boise, Idaho*

EDITOR'S REPLY: Thanks, Lizbeth. Watch for Mini Maker Faires in Kitsap County, Wash., and Vancouver, B.C. Or why not organize one? See makezine.com/go/makeamakerfaire.

WikiLeaks' persecution ("Make Free," Volume 26) and desperate despots in the Middle East shutting down the internet are a warning! Legislating net neutrality is not enough. The onion router and distributed-computing "leak-server" apps are not enough.

Freedom-loving makers need to own their own network infrastructure. I'd like to see how-tos on building long- and short-range data distribution systems — especially optical networks, because they minimize RF interference and get more bandwidth per watt on a maker's power budget. Could optical audio cables carry network packets? Would a $200 telescope lengthen a Ronja's range?

—*Evanie Cronquist, Skull Valley, Ariz.*

MAKE AMENDS

In Volume 26's *Flame Tube* Tools list, page 75, the drill bits should be 1/16" and 9/16".

Volume 26's Make Amends incorrectly corrected the equation values in Volume 25's "Remaking History." Aargh! The crank length (a radius) is 6" as originally stated, but the radius of the large barrel is 1½". The value of the expression is therefore 96, not 48.

MAKER'S CALENDAR
Compiled by William Gurstelle

Our favorite events from around the world.

World Maker Faire New York
Sept. 17–18, Queens, N.Y.

Last year saw the debut of World Maker Faire New York, the biggest maker-oriented event on the East Coast. From loud, fiery pulsejets to Tesla coil lightning to Arduino-controlled sculpture to some of the best technology presentations of the year, there was something for everybody. This year promises to be even bigger and better.
makerfaire.com/newyork/2011

AUGUST

⌃ Les Cheneaux Islands Antique Wooden Boat Show
Aug. 13, Hessel, Mich.
An old joke says the happiest day of a boat owner's life is the day she buys her boat, and the second happiest is the day she sells it. But wooden boat enthusiasts love working on their boats and showing them off. This big meet includes dinghies, rowboats, canoes, launches, sailboats, runabouts, and cruisers. lchistorical.org/boatfest.html

» Magic Live!
Aug. 14–17, Las Vegas
Magic and making go hand in hand — so much so that we devoted a MAKE issue to it (Volume 13). Tricks, illusions, and props require a lot of secret, one-of-a-kind constructions. One of the biggest magician gatherings is this "unconventional convention" in Las Vegas (where else?). Here, the best and brightest will demonstrate some of their top tricks, and perhaps share some of their best-kept secrets.
magicmagazine.com/live

SEPTEMBER

» Festival of the Building Arts
Sept. 17, Washington, D.C.
This festival celebrates the past and envisions the future of construction. Nascent builders can design their own buildings from recycled craft materials to create a City of the Future. Visitors of all ages can try their hand at drywall finishing, plastering, roof thatching, woodworking, surveying, nail driving, and more. nbm.org/families-kids/festivals/foba.html

» The Pacific Pinball Exposition 2011
Sept. 23–25, San Rafael, Calif.
Welcome to the electromechanical underground — the world of light, sound, relays, switches, and solenoids that we visited back in MAKE Volume 08. The PPE boasts a huge variety of pinball and other games as well as historical displays, unique artwork, and science exhibits. pacificpinball.org/events/ppe-5

» U.S. Department of Energy Solar Decathlon
Sept. 23–Oct. 2, Washington, D.C.
When President Obama — "the Maker President" — throws down the gauntlet, it's hard to resist answering the challenge. Twenty competing university teams come to our nation's capital to design, build, and operate solar-powered houses. The results of the contest will stand proudly on the National Mall.
solardecathlon.gov

OCTOBER

» Great Western War XIV
Oct. 5–10, Taft, Calif.
Catapult warrior, hast thou been looking for a good place to launch thy weapon? Then consider a journey to the Great Western War, a major gathering of the Society for Creative Anachronism. The event features a great number of ancient science- and crafts-related events. caid-gww.org

» Steamcon III
Oct. 14–16, Seattle
The world of Victorian science fiction — airships, submersibles, rayguns, and mad scientists — is brought to life when aficionados of steampunk (highlighted in MAKE Volume 17) assemble. Steamcon is a great event for those who love brass gears, goggles, and Jules Verne-like machines — this year's theme is *20,000 Leagues Under the Sea*. steamcon.org

IMPORTANT: Times, dates, locations, and events are subject to change, Verify all information before making plans to attend.

MORE MAKER EVENTS:
Visit makezine.com/events to find classes, exhibitions, fairs, and more. Log in to add your events, or email them to events@makezine.com. Attended a great event? Talk about it at forums.makezine.com.

IN THE MAKER SHED
By Dan Woods

Getting Started:
3,100 New Electronics Makers

OUR 6TH ANNUAL BAY AREA MAKER FAIRE
drew to a close just 72 hours ago, and I'm
working through the post-Maker Faire blues
— transitioning from the adrenaline rush of
being surrounded by 100,000 passionate DIY
enthusiasts one day to the relative serenity of
the office the next. It's enormously satisfying,
however, to reflect on the profound impact
that Maker Faire has on attendees.

The heart of the Maker Shed's mission is
to help people get started making. Through
project-based kits like Getting Started with
Arduino and Getting Started with Compressed
Air Rockets, we help newcomers start making
for the first time and help experienced makers
learn new skills.

Nowhere is this more true than in the Maker
Shed pop-up stores we operate during Maker
Faire: 14,000 square feet and hundreds of
kits and components combined with dozens
of makers, authors, and volunteers helping
people explore everything from Arduinos to
cheese making.

And nowhere is our mission more visible

**Learn to Solder »
Merit Badge Kit**
$3, Product code MKLSOL
makershed.com/pin11

than in the Maker Shed's "Learn to Solder"
tent where, with the helpful coaching of Mitch
Altman, Jimmie Rodgers, and a dozen local
hackerspace volunteers, some 3,100 attendees mastered soldering for the first time.
Participants left the Maker Shed with smiles
on their faces, new skills, increased confi-
dence, and a cool flashing LED merit badge to
show for it. The perfect gateway project kit.

If you'd like your own Learn to Solder merit
badge kit — or you're looking for a cool gateway project or for a group or after-school
class — they're available for purchase in the
Maker Shed (makershed.com/pin11). We'll
also include one with every Maker Shed order
of at least $10 through August 31. It's on me. ▨

Dan Woods is MAKE's associate publisher and general
manager of e-commerce.

Face to Face

Tom Banwell is a self-taught man of many talents. He's a leatherworker, a caster/sculptor, and a tireless inventor of a vast selection of imaginative facemasks, many of which have been featured in films, television, and major magazines.

His most complex and extraordinary works are his "steampunk" gas masks, but he's also known for his delicate, laser-cut leather party masks and other uniquely shaped costume masks. Just to keep things interesting, he also makes rayguns.

His fantastic blog is a must-read for any costume designer or lover of steampunk. It's filled with well-written, step-by-step explanations and interesting tips and tricks. (Be sure to search for "A Steamier Raygun Holster," "Elevated Shoes," and "Modifying a Straw Hat.")

When asked why he gravitated to gas masks, Banwell says, "A gas mask, though functional, dramatically alters the appearance of the wearer. This can be perceived by the viewer as terrifying — as one resembles a monster — or humorous — as one becomes a silly clown."

Banwell manages to combine these two feelings to create unforgettable masks that embody both fear and curiosity. The formal, antiqued leatherwork feels classic and fore-boding, but he says the form of the masks — which can resemble a rhinoceros or an elephant — is "pure fantasy."

Banwell is constantly looking at the world around him and re-creating it in the most mad and pleasing manner possible. Looking through his fan photos, it's clear that when seemingly ordinary people don his masks, they unleash the more fantastic selves that lay dormant.

—*Stacey Ransom*

» **Steamy Leather:** tombanwell.blogspot.com

Tom Banwell

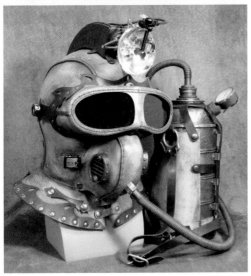

Analog Wranglers

Brian Dewan may be a jack-of-all-trades, but he's clearly a master of many. An inventor and builder, this Catskill, N.Y., resident is also a performance artist, cartoonist, and multi-talented musician.

Dewan plays and sings along with semi-traditional instruments like autoharps, zithers, and accordions, making them seem appropriate for everything from heavy metal to folk songs. He also performs with the cheerfully lunatic Raymond Scott Orchestrette. (Raymond Scott composed the music for numerous Warner Brothers cartoons.)

In collaboration with cousin **Leon Dewan**, Brian has created a series of instruments called Dewanatrons that vary in complexity from simple, elegant, wall-mounted interactive sound sculptures such as their *Melody Gins* and *Courtesy Modules* to the extraordinary *Dual Primate Console Mark II*, shown here.

It's impossible to precisely control, so the "primates-in-charge" must guide the machine to produce melodies, percussion, and innumerable unpredictable sound effects. Brian explains, "It is the responsibility of each primate to encourage or restrain the instrument."

Some simpler pieces are best experienced in unison: a dozen *Wall Gins*, for instance, were displayed in Brooklyn's Pierogi Gallery, configured to play simultaneously, creating an eerie and enthralling ambient soundscape.

All the musical instruments are "analog, solid-state" works covered in knobs and dials, full of oscillators and other physical sound manipulation controllers, with beautifully finished cases that bespeak Brian's fine-furniture-building background. They are artifacts and heirlooms of a Dewan-ized version of the past.

—*Rose White*

» Dewan's Instruments: dewanatron.com

Joshua Sarner

Turn Signals and Turntables

Combine two unrelated things inside the brain of Dutch artist **Olaf Mooij** — the electronica song "God is a DJ" by Faithless, and the Popemobile — and the result is the DJ Mobile.

With 14 woofers, tweeters, and other loudspeakers boldly splaying in various directions out the back of a dark blue 1983 Ford Sierra, the DJ Mobile has a surreal, cartoonish look.

Not only does the car have a professional-quality Beyma sound system, but it's also got a DJ setup, with a place to set up turntables and a mixer on the roof of the car. There's a hole in the roof as well, where the DJ can stand and mix.

The idea came to Mooij, 53, when he heard the song and then thought of the infamous vehicle that the Pope makes public appearances in — a normal-looking car in front with a bulletproof glass, pope-sized display case in the back.

Mooij bought the used car from a local gas station. "I always liked the Ford Sierra," he says. He worked up the quirky design using buckets, cardboard, and duct tape. Later, he swapped the prototype materials out for wood and polyester.

As you might imagine, the DJ Mobile commands a good deal of attention when it gets driven around. Though it's street legal, Mooij notes that driving around can be dangerous because of how distracting it is to other drivers. The people most noticeably affected when in the presence of the DJ Mobile are "the guys with big car hi-fi sound systems," says Mooij, who get "pale with jealousy."

Next up: Mooij is currently building a temple out of three old Volkswagen Beetles.
—*Laura Cochrane*

» **DJ Mobile:** olafmooij.com

Hutt's Cantina

When **Jason Hutt** was just 2 years old, his mother took him to see *Star Wars* at the movie theater. He sat in rapt silence throughout the entire film, and a lifelong obsession was born.

"There's really never been a time in my life when I didn't love *Star Wars*," he says. "My mom loved space and science fiction; she cultivated that interest in me throughout my childhood, and *Star Wars* was really the center of it."

As Hutt's collection of *Star Wars* action figures (1,600 and counting) and memorabilia grew, he realized he needed a fun way to display it. His penchant for DIY projects led him to creating a massive and intricate diorama of the Mos Eisley cantina scene.

"The cantina scene has to be one of the most iconic in all of science fiction," he says. "It's where the adventure truly begins as the heroes meet for the first time and begin their epic journey."

By day, Hutt works in NASA's International Space Station Mission Control at Johnson Space Center, and is father to three young girls. It was by night, after his daughters went to bed, that his project took shape. Working for about an hour each evening, it took nearly four months to complete.

He started with sketches and lots of research and chose MDF to create the base and walls of the structure. From there, he had to get creative to find ways to replicate some of the details of the iconic scene. The bar posed the greatest design challenge.

"The bar was a mini-project in itself, as I used a combination of wood spheres, dollhouse pipes and accessories, some empty jars, and a half dozen other things," Hutt recalls.

Now that his cantina diorama is complete, Hutt is beginning work on Jabba the Hutt's Throne Room and also has plans to create an Endor bunker-fight diorama.

—*Rachel Hobson*

Drill-Powered Future Trike

In MAKE Volume 26, we featured a drill-powered go-kart you can build at home. As cool as that vehicle is, it looks positively retro compared to the EX, a futuristic drill-powered vehicle designed by promising young German designer **Nils Ferber**.

Ferber's work explores outside-the-box thinking and the potential of design to "alter reality," he writes. The EX (a play on *eccentric*, as in "deviation from what is ordinary or customary") sure meets these criteria, with its sleek, *Tron*-worthy shape and its complete rethinking of a steering mechanism.

Working with fellow designer/builders **Sebastian Auray**, **Ruben Faber**, and **Ludolf von Oldershausen**, Ferber and team started out by prototyping designs in Lego blocks, wood scraps, and finally steel. The final trike is fashioned in stainless with many CNC and specially fabricated parts.

Through their design process, they developed a unique way of driving and steering the vehicle. The driver lies down on the EX, forward-facing. This creates "an exciting driving experience" and allows the driver to operate the strange "spine-shaped joint" steering.

To steer the EX, the driver has to employ body weight to flex the six axles of the basically groin-mounted spine-joint, a sort of short spinal column that's used to flex and bend the vehicle in the desired direction of travel.

Propelled by two Bosch 18-volt cordless drill/drivers, the EX achieves speeds of 30 kilometers per hour (almost 19mph). And given the position of the driver and proximity to the ground, we can only image the thrill ride offered by this ingenious crotch-rocket. Let's just hope that EX doesn't take on any unintended additional meanings.

—*Gareth Branwyn*

» **EX Riders:** nilsferber.de/ex.html

Pedal Power

As gas prices soar, **Ian Fardoe** of Staffordshire, England, is faring better than most — at least better than those of us who drive cars. The 40-year-old is car-free and always has been.

"I've been cycling all my life. I did learn to drive a car but realized that I hated it a long time before I took my driving test, so didn't even bother," he explains. And with the price of petrol above 130 pence a liter (about $8 a gallon) this year, he's much better off.

Fardoe commutes 4,000 miles a year by bike, so building a wacky, plastic-wrapped tricycle is no surprise. Called a *velomobile*, the trike's outer shell is made from repurposed corrugated plastic that was harvested from a recumbent bike's fairing, and fashioned after vehicles found in Australia's Pedal Prix.

The first build took about six weeks, but Fardoe spent 18 more months tweaking the design to make it more practical and efficient. The initial frame weighed in at only 11lbs and

didn't stand up to his 6-mile commute to work.

"I bought some hot glue and zip ties, etc., to make it look remotely pretty. The making it pretty bit didn't work very well," he jokes.

Overall, the vehicle cost about £2,600 ($4,000) to build — far less than commercial velomobiles, which can cost twice that price.

Fardoe admits his way of building is very learn-by-doing, which is how his vehicle earned the name *OTP (On the Piss)*.

But hot glue and zip ties aside, it's a pretty serious ride. He's done 125 miles in a single trip and reached a top speed of 73mph going downhill, 53mph on flat ground.

In the process, Fardoe has become a pretty serious velo pilot; at press time he was planning to race a commercial velo in the Cycle Vision races in the Netherlands in June.

—Jerry James Stone

» **Fardoe's Velos:** picasaweb.google.com/ian.fardoe

Ian Fardoe

Pendulum Perfector

If you've visited enough science museums, you've seen a Foucault pendulum. Named after its inventor, French physicist Léon Foucault, the pendulum demonstrates the Earth's rotation by knocking down pins arrayed around the pit over which it swings. Although the pendulum appears to swing around the pit, it's the pit, and Earth it stands on, that rotate beneath the pendulum. And there's a good chance the pendulum you know is one of more than 100 made so far by **Cary Ponchione**.

For 34 years, Ponchione worked in the basement shop of the California Academy of Sciences museum in San Francisco. Shortly before moving from its home of 87 years in Golden Gate Park, the Academy downsized its in-house exhibit-building shop and offered early retirement to Ponchione.

But calls for the pendulums that the Academy could no longer make kept coming

in. "They asked me to take [them]," Ponchione says of the pendulum orders, and a small business, Academy Pendulum Sales, was born.

A friendly, outgoing machinist and fabricator, Ponchione keeps a spare, tidy shop by the tracks in Richmond, Calif., where he assembles pendulum kits for customers to install. "As you can imagine," the semi-retired maker deadpans, "it's not a full-time job."

Though he has others join, turn, finish, and polish the cast brass hemispheres that make the 235-pound, 16-inch bob, Ponchione makes and assembles most of the small parts, and still hand-winds the ring electromagnet, which adds the perfectly timed kick that keeps each pendulum swinging.

—*William Abernathy*

›› **Foucault Pendulums:** calacademy.org/products/pendulum

TALES FROM THE WEB
By Gareth Branwyn

Building a MAKE Video Network

WE'VE GREATLY EXPANDED MAKE'S VIDEO offerings with a regular roster of excellent project and tutorial series, all conveniently accessed from blog.makezine.com/video and youtube.com/makemagazine and iTunes. We've come a long way since our original Weekend Projects series! Tune in to see what we've got going on.

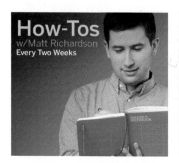

☼ **How-Tos with Matt Richardson** In this biweekly series, Matt shows how to make a cool project, from a physical mailbox that tweets your mail's arrival to fun photography hacks to an Arduino-powered music box. blog.makezine.com/tag/howtoswithmatt

☼ **Becky's Workshop** Every month, our CRAFT/MAKE crossover star, Becky Stern, presents an innovative project to add some high-tech high-touch to your life, from pocket reflectors for cycling to T-shirt hacks to working with electroluminescent wire. blog. craftzine.com/archive/video_beckys_workshop

☼ **Collin's Lab** Every two weeks, Collin Cunningham presents a new project or explains an electronics concept in entertaining and easy-to-understand language. blog.makezine.com/tag/collinslab

☼ **The Latest in Hobby Robotics** Tune in weekly as Frits Lyneborg and friends from the Let's Make Robots community discuss the week in DIY robots. blog.makezine.com/tag/latestinrobots

☼ **Make: Live** Our increasingly popular streaming show and tell, hosted by Becky Stern and Matt Richardson, airs every second and fourth Wednesday of the month. Watch as they bring MAKE magazine to life, with in-studio and on-cam makers showing off their projects. makezine.com/live

☼ **Tiny Yellow House** This show from prolific backyard builder Derek "Deek" Diedricksen has been described as "*This Old House* meets *Wayne's World*." Each month Deek builds forts, cabins, and "microhomes," and teaches woodworking, construction, reuse, and recycling in the process. blog.makezine.com/tag/tinyyellowhouse

And that's not all. We've also got a weekly **Meet the Makers** series, **Super Awesome Sylvia's Mini Maker Show**, **Home Décor with Meg Allan Cole**, **Corinne's Craft Closet**, **The Latest in Arduino**, and more. Check the MAKE video page (blog.makezine.com/video) and watch them all. ▰

Gareth Branwyn is editor-in-chief of makezine.com.

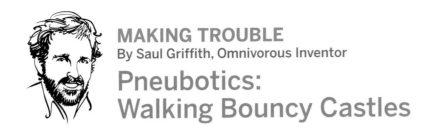

MAKING TROUBLE
By Saul Griffith, Omnivorous Inventor

Pneubotics:
Walking Bouncy Castles

SOMETIMES I FEEL LIKE A FALSE NERD, or a geek with two important genes missing: I'm not particularly interested in space exploration, except as fiction, and I've never cared for robots. So I find it strange that I'm now working on a Defense Advanced Research Project Agency (DARPA) robotics program.

I think what I never liked about robots is that they're complex machines that really don't do much. They're fragile and very expensive. I like simple, robust things; things that don't cost more than they should.

What I've found myself working on (with Jack Bachrach, Geoffrey Irving, Pete Lynn, and the good guys from Meka Robotics) is completely soft, completely compliant, very lightweight, and very cheap. No joints. No servos. Just skins — inflated skins.

For a long while I've been fascinated by inflatable objects for their extreme strength-to-weight ratios (they can carry a lot of load for very little mass). I also love the challenge of designing something "human safe," in the robotics lexicon. Biology doesn't use metal, and it doesn't use servos. Nature points to some very interesting alternatives.

To make it work, we had to invent a new kind of actuator. Think of it as a vascular system for robots. It's fluidic — works equally well with air or water — and by pumping either of those around, you can change the dimensions of the skin and effect motion. Our first actuator was quite literally a bicycle inner tube in a sewn pair of membranes. It worked really well for a $5 prototype!

For the next trial, I asked my sister to return an inflatable 4-foot-high elephant I'd designed and given to my niece. When it arrived, Pete burned the midnight oil and sewed up some vascular "muscles," and in a day or two we had four moving legs. It actually walked. About one mile every 24 hours, but hey — baby's first steps! It moves like no machine you've ever seen; more like the way biology moves. A walking inflatable elephant might sound ridiculous, but it works, and the numbers on paper told us it should have incredible strength, good speed, extremely low weight, and cost very, very little to manufacture.

> Biology doesn't use metal, and it doesn't use servos. Nature points to some very interesting alternatives.

The next prototype was designed to walk with a human rider on it and to look less like an elephant. We built it in under a week for less than $1,000 in parts. A 15-foot-long, 5-foot-high robot with 28 muscle actuators (four in each of six legs, another four in the trunk). It worked too (after a few exploded actuators).

I like the idea of a robot you can sew together. I like that it has no heavy, sharp, or costly parts. Most of all, I like the intellectual challenges of it. There aren't any CAD packages for designing highly elastic kinetic membrane structures. We had to write our own. There aren't any analysis simulations. We had to write our own. There aren't any walking bouncy castles out there. We built our own! We call our weird new style of robotics "pneubotics," as in *pneu* for air (like pneumatic).

Who knows if the robotics community will like it or even care. Either way, that's not why I built it. I built it because perhaps my niece will forgive me if she gets a walking elephant next Christmas that she can ride to school. ◪

Saul Griffith is chief troublemaker at otherlab.com.

COUNTRY SCIENTIST
By Forrest M. Mims, Amateur Scientist

Startups: Origins of the PC Revolution

TODAY'S SMARTPHONES AND TABLETS, laptop and desktop computers all trace their ancestry to the arrival of the hobby computer era of the 1970s.

After Intel introduced its 8008 microprocessor in 1972, several individuals and teams began using the new chip to build DIY computers. But these computers made little progress due to the 8008's limited capabilities.

The computer revolution was jump-started in 1975 when MITS, Inc., a small electronics company that I co-founded in Albuquerque, N.M., announced the Altair 8800, a kit computer designed around Intel's new and powerful 8080 microprocessor.

Many books have been written about what happened next, and *Idea Man* (Portfolio/Penguin), a new memoir by Microsoft co-founder Paul Allen, shines a spotlight on many details that were previously known only to insiders. Whatever your favorite kind of computing device or operating system, *Idea Man* is a book well worth reading, especially if you have entrepreneurial aspirations.

The story begins at Out of Town News in Cambridge's Harvard Square on a snowy December afternoon in 1974. Allen visited the newsstand each month to check out the latest issues of *Radio-Electronics*, *Popular Science*, and similar magazines.

When he saw the January 1975 issue of *Popular Electronics*, it stopped him in his tracks. Emblazoned on the cover was a photograph of the Altair 8800 microcomputer. The blurb over the photo read:

> PROJECT BREAKTHROUGH!
> World's First Minicomputer Kit
> to Rival Commercial Models …
> "ALTAIR 8800" SAVE OVER $1,000

Allen opened the magazine and found complete construction plans for the Altair 8800, which was available as a kit ($439) or fully assembled ($621). He noticed that the core of the Altair was Intel's powerful new 8080 microprocessor, the successor to the 8008. He paid 75 cents for the magazine and hurriedly strode almost a mile to Harvard's Currier House, where sophomore Bill Gates resided.

Gates shared Allen's enthusiasm for the Altair. Both had become expert assembly language programmers in high school, and they decided to contact Altair developer Ed Roberts, who headed MITS, Inc. Their plan was simple: offer Roberts a version of the BASIC language that would run on the Altair.

After eight grueling weeks of programming, Allen flew to Albuquerque with a paper tape, punched with their new BASIC. The code ran fine when simulated on a PDP-10 minicomputer at Harvard, but would it work with an Altair? While Roberts watched, Allen carefully entered into the Altair's front panel toggle switches the code he'd written on the airplane to enable the Altair to load the BASIC from the Teletype terminal connected to the computer. The paper tape reader then loaded the BASIC into the Altair's memory. When Allen typed PRINT 2+2, the Teletype immediately printed 4.

Roberts was amazed. So was Allen, though he didn't let on. Soon Roberts hired Allen, and later that year Gates joined him in Albuquerque. There, Allen and Gates formed a partnership that they initially called Micro-Soft.

Idea Man

Allen tells what happened next in *Idea Man*, a detailed and appropriately technical account of the origin and early history of Microsoft. It's much more than a book about microcomputer history and Allen's life as a billionaire,

for it's packed between the lines with tips for aspiring entrepreneurs, designers, programmers, and makers with revolutionary ideas.

Idea Man has attracted considerable attention in the media world because of its candid revelations about friction between Allen and Gates and what Allen describes as Gates' efforts to reduce Allen's stake in Microsoft.

The shouting matches he describes closely parallel what Roberts and others told me over the years. Some believe that dredging up these old stories is sour grapes, especially since Allen played much less of a role at Microsoft after his 1982 bout with cancer and his growing disillusionment with Gates' confrontational leadership style.

Having just spent four years writing an exhaustive history of the world's leading atmospheric monitoring station, Hawaii's Mauna Loa Observatory, I disagree. Debates, arguments, and leadership flaws, whether in the low-pressure environment of a remote station at 11,200 feet or in the high-pressure environment of a startup company, are the sparks that illuminate the organization's

history. Allen would have short-changed his readers had he failed to describe the disputes.

Allen even describes an expletive-laden temper salvo directed by Steve Jobs against a hapless Apple employee while he and Gates watched with surprise. Leadership antics like these will provide business analysts, academics and, yes, psychologists much to ponder when they study the astonishing success of Microsoft and Apple.

Whether these disclosures have burnt the bridge in the four-decade relationship between Allen and Gates remains to be seen. In January 2011, four months before the release of *Idea Man*, Allen was in Albuquerque to dedicate Startup, a personal computer museum gallery, to the memory of Ed Roberts, who died in April 2010. When I asked Allen about his book, he said he was concerned how Gates would react.

Gates seems to have mellowed over the years. After he joined Allen in Albuquerque in the mid-70s, the teenage-looking Gates sometimes had major battles with the burly Roberts, a former Air Force officer who expected respect. Last year when Gates learned that Roberts was near death, he flew across the country to spend several hours with him days before Roberts died.

Allen writes in *Idea Man* that Gates regularly visited him in 2009 when he was hospitalized with his second battle with cancer: "He was everything you'd want from a friend, caring and concerned." Based on their past history, it seems likely the two billionaires will eventually make their peace, perhaps while agreeing to disagree on some points. After all, many Microsoft customers who have a love-hate relationship with the company's software (including me) keep going back for more.

After Microsoft

Microsoft made its founders two of the world's richest men, and *Idea Man* follows Allen's account of the MITS-Microsoft years with highlights about his life, business, and philanthropy. He enthusiastically discusses his billionaire lifestyle, including his sports teams, his love affair with the guitar, and his far-flung travel adventures aboard his mega-yachts.

Much more important to us makers than the celebrity name dropping and travel stories are the details of Allen's business successes and failures, his founding of the Allen Institute for Brain Science, and his carefully restored World War II-era aircraft collection.

Then there's Allen's partnership with Burt Rutan that culminated in *SpaceShipOne*, the first privately developed and launched reusable, manned spacecraft. The historic *SpaceShipOne*, which earned the $10 million Ansari X Prize, is now suspended between Lindbergh's *Spirit of St. Louis* and Chuck Yeager's Bell X-1 at the Smithsonian Air and Space Museum in Washington, DC.

Lessons for Makers

Idea Man provides important tips and lessons for today's generation of makers, some of whom might even now be developing what might become the next billion-dollar technology or product. Here are some lessons I've gleaned from its pages and between the lines.

» Does your idea pass the balloon test? Good ideas and futuristic visions don't guarantee successful products and ventures. As Allen wrote about his pre-Altair days with Gates, "Each time I brought an idea to Bill, he would pop my balloon."

» Texas-style handshake agreements with partners, supporters, and customers are great. I sold millions of books to RadioShack over handshakes and purchase orders. But Allen's experience suggests it's best to follow handshakes with carefully drafted agreements that all concerned are willing to sign.

» Use care and prudence when working and dealing with partners and financial backers.

» Get to know your partners and their idiosyncrasies before signing on with them.

» Carefully read any agreement or contract before you sign it!

» Partnerships are a two-way arrangement. So get to know yourself. Are you living up to your agreements? Is your management style reasonable or do you create chaos?

» A partnership agreement should provide contingencies for all eventualities. For example, the partners should agree to pursue arbitration in the event of a serious disagreement. The agreement should cover what happens should a partner be incapacitated or die.

» Never, never, never release imperfect products! Delaying a promised new product is always better than releasing a defective one.

» Treat your customers with the respect they deserve.

» As Roberts learned so well, if your first products don't succeed, try again. ▨

Forrest M. Mims III (forrestmims.org), an amateur scientist and Rolex Award winner, was named by *Discover* magazine as one of the "50 Best Brains in Science." He was a co-founder of MITS, Inc. and wrote the first Altair 8800 user's guide.

MAKE FREE
By Cory Doctorow, Digital Rights Defender

Four Horsemen of the 3D Printing Apocalypse

CHARLES STROSS' EXCELLENT NEW novel, *Rule 34* (Ace Books), is a futuristic police procedural set in a near-future Edinburgh, in which 3D printing has become boringly ubiquitous.

You can buy safe, prepackaged 3D printers at the local housewares shop, and they're handy for whipping up generic replacement parts for broken appliances (at one point a character drops and cracks his wife's cherished German onion slicer and realizes he'll have to google for a 3D file to match the broken piece) or paid-for 3D files licensed from big media companies.

These printers are controlled by DRM that checks jobs against a blacklist of forbidden shapes and prevents them from being output if they match (presumably there's some loose matching algorithm in use that can get past ruses like simple changes to the shape).

Of course, this doesn't work. The bad guys handily outmaneuver the prohibitionists, and a black market springs up, producing such wicked marvels as solid-state meth labs, brass knuckles made from super-hard polymers, and X-ray-invisible, nonferrous stick-up knives.

I think Stross' speculation on the future of 3D printing gets the shape right, even if the details might not match exactly. Since the early days of computer regulation, hysterics have made recourse to the "Four Horsemen of the Infocalypse": child pornographers, organized crime, terrorists, and pirates. Invoking one or more of these terrible fellows is often sufficient to stifle further debate and end critical thought ("Won't someone think of the children?!").

It's not that bad guys don't use our beloved machines to do bad things. But a prohibition against running certain programs is a nonstarter. In practice, a computer's owner can, with sufficient technical knowledge (or access to a large, searchable database of general knowledge, such as the internet), trivially unlock her device so that it can execute any valid program.

But when all you have is a hammer, everything starts to look like a nail. In Stross' world, as in our own, the regulatory response is to build devices that have internal snitches that check to see if their owners are running naughty unlocking programs. And in Stross' world, as in our own, the need to prevent the dissemination of snitchware countermeasures leads to widespread surveillance and censorship of the internet.

> Pro-regulation hysterics make recourse to the Four Horsemen of the Infocalypse: child porn, organized crime, terrorists, and pirates.

And in Stross' world, as in ours, none of this actually works worth a damn at stopping bad guys. Instead, it creates a vicious cycle of more surveillance and more control to overcome the failings of the current round of censorship and surveillance.

Rule 34 abounds with imaginative horrors about the potential for technology to do bad, and as imaginative as Stross is, I'm guessing that the real thing will be even ickier.

We need real solutions to the bad stuff that people come up with when they get technology. The first step to finding a real solution is to stop doing things that don't work. ◪

Cory Doctorow's latest novel is *Makers* (Tor Books U.S., HarperVoyager U.K.). He lives in London and co-edits the website Boing Boing.

Maker

GLENN DERRY: MOVIE MAKER

Avatar's special effects innovator hacks together blockbuster filmmaking tools — and shares his recipe for an indie-budget virtual camera.

By Bob Parks

When the credits finally scroll up after a great special effects movie, don't you wonder at the alphabet soup of job titles on the screen? (What's the difference between a previs supervisor and a virtual production supervisor?) Well, regardless of title, there are always a few key engineers who work at a director's side, solving near-impossible problems without complaint. When the director asks for a 2-ton *Tyrannosaurus* that moves like a ballerina, these guys are on it. When the director asks for a new way to film giant blue people flying through the air, they say, "No problem."

Glenn Derry is one of those guys. The 36-year-old owner of a small Hollywood engineering company called Technoprops spends his days solving specialized challenges on the set. When he was 16, he worked for director Steven Spielberg on *Jurassic Park*, making a dinosaur move gracefully. At 30, he worked on *Avatar*, taking orders from James Cameron to create

a "virtual camera" device that lets directors climb inside computer-generated films as they shoot real-life actors (see page 37 for an indie film version you can build).

The virtual camera is so effective that it's becoming de rigueur on effects films, necessitating Derry's travel to movie sets — including those of *Real Steel*, in which Hugh Jackman coaches robots in arena boxing, and *The Adventures of Tintin*, Spielberg's motion-capture animated version of the classic comic (both due in theaters later this year).

By now, Derry's almost used to the frantic pace of technology development in the movie business. "Nothing is ever carefully planned out," he says. "Your job is to solve problems fast and hope to God you don't hurt anyone with a dinosaur tail bigger than a Kubota backhoe."

Bob Parks is a frequent contributor to MAKE, *Runner's World*, and *Wired*. He lives in Vermont with his wife and two children. He can be reached at xbobparksx.com.

Noah Webb

VIRTUAL VIRTUOSO: Hollywood F/X innovator Glenn Derry perches on racks of equipment to be used in future CG film endeavors.

THE DAY I SPOKE WITH DERRY, HE WAS chatty and in good spirits, though physically exhausted. Three days earlier he'd had neck surgery to remedy a pinched nerve, but he denied any connection between the neck brace he was wearing and his demanding schedule of maintaining his cameras and using them on film sets around the country.

All he could talk about were the fun side projects he has on deck. In one corner of his 18,000-square-foot prop fabrication shop stands a hydraulic lift that once moved the robotic armor "AMP suit" in *Avatar* but now might be modified into a promising Formula 1 simulator. In his home workshop are remnants of giant hydraulic chimes and gongs he constructed to make a colossal musical instrument.

Elsewhere, Derry's staff of ten mechanical and electrical engineers and programmers tinkered with an open source CNC mill as it churned out metal parts for an upcoming film production. By hacking around like this, Derry and crew have cultivated the uncanny ability to chew up random consumer products, bits of castoff metal, and fiber-optic cables, then spit it all out in the form of tools that have become essential to modern spectacle filmmaking.

Derry's father worked in the film industry as a welder, machinist, and physical effects guy ("From the time I was a teenager, my dad blew up cars," he says), and this connection and some lucky timing landed him an internship on *Jurassic Park*. The need for good engineers became clear to Derry his first time on the set, where the teenager worked on giant robotic dinosaurs (the film used physical puppets as well as computer-generated dinos). "I was soldering op-amp circuits by hand, one at a time," he says.

But the production soon ran into major problems. When you move a robot bigger than a carnival ride, the great mass of steel shakes and sways awkwardly. Already a budding electronics whiz, Derry helped come up with the solution: mount accelerometers on the dinosaurs' limbs and feed the signals back into the controller. Making the dinosaurs move more gracefully turned out to be a formative experience, recalls Derry. "I was 16, surrounded by electronics card cages and getting screamed at by Spielberg and Stan Winston."

He started college, studying to be a musician, but the lure of movie work was too much. (Years later he returned to study mechanical and electrical engineering at UCLA.) Derry worked full time as a puppeteer on the second *Jurassic Park* film in 1997, where they started out using the same animatronic technology they'd used to control the shark head in *Jaws*.

But Derry, who'd spent years in his bedroom sequencing keyboard music with MIDI controls, put two and two together: he rigged up a MIDI interface that would let any desktop music package control a dino's facial gestures and fine movements, resulting in more credible movie monsters.

All this prepared him for the toughest challenge yet of his career: *Avatar*. To make that film, Cameron became fixated on a motion capture system that would track the actors' faces as well as their bodies, a feat never before achieved in film production. Derry, the virtual production supervisor, found a sensor that was simple and reliable enough to attach to a little boom in front of the actors' faces.

Cameron also wanted to direct scenes that mixed live actors and animated characters. But motion capture systems couldn't be used near live-action filming because the little reflective dots on motion capture suits don't show up under white-hot studio lights. Derry told Cameron it was an easy fix, but he labored two years to solve it. He and his team built special LED trackers that blinked at the exact frame rate of the motion capture cameras, which enabled the cameras to see them.

> **❝** I was 16, surrounded by electronics card cages and getting screamed at by Spielberg and Stan Winston."

⚲ MECHA MOTION: Derry's 7-axis motion controller (top left) is at ¼ scale to the full-sized hydraulic 7-axis motion base (bottom left) used on *Avatar* to move the massive, robotic AMP suit torso. His sprawling prop fabrication shop (top right) is overseen by plaster casts of actors' faces from various projects (see if you can spot Daniel Craig, Simon Pegg, Nick Frost, Andy Serkis, Cary Elwes, and Timothy Dalton), digitally scanned and 3D-printed.

Cameron wanted to live and breathe inside Pandora, the jungle moon setting for *Avatar*, and it was Derry's job to make it work. It started when visual effects expert Rob Legato suggested the director use a system that would let him look inside the artificial world of the movie, frame shots, pick lenses, and choreograph camera movements. Derry built a device that resembled a 16mm movie camera with an LCD eyepiece. The idea was that the director could carry it around the set, checking out different angles, but the camera shape was cumbersome. It was then that Derry developed the virtual camera, a 7-pound package that automatically swings toward

✎ **PANDORA BOX:** (Left) Director James Cameron used Derry's virtual camera on the set of *Avatar* to live, breathe, and compose shots inside the world of Pandora. (Right) Derry's facial capture head rig from *Avatar* transformed Zoe Saldana's performance into an incredibly credible 9-foot-tall blue warrior princess.

the user no matter how it's held.

Directors have delighted in the way the technology connects them artistically with moviemaking again. Spielberg told Derry that using the virtual camera felt like shooting Super 8 film in the backyard as a kid: "He said *Tintin* felt handmade, even though there was this giant juggernaut of a special effects company behind him."

A product so valuable to the movie industry would seem to be easy to commercialize. Instead, Derry has been hiring himself out, before rival effects experts figure out how to make their own virtual cameras. And Cameron insists that the concept be shared with other directors. "If Jim wants to share it, I'm not going to tell him he can't," Derry shrugs. "Besides, I was able to prototype stuff on the most expensive movie ever made, so now I can bring the technology to smaller-budget films."

But with *Avatar 2* in production, Derry's busy schedule precludes indie filmmaking or other hobbies. Derry looks wistfully at his old electric bass and piano in his Santa Clarita, Calif., home, which he shares with his wife and two kids. "As many would-be rock stars learn," he says, "you eventually get a real job." Although making 9-foot blue people fly around isn't exactly a day at the office for most of us. ◪

➕ See makeprojects.com/v/27 for a video profile, slideshow, and virtual camera how-to.

Make Your Own Virtual Camera

Learn how at makeprojects.com/v/27.

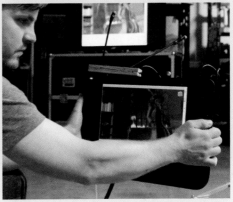

Noah Webb

"THERE DOESN'T HAVE TO BE A CRAFT services truck out front to make a high-quality computer-generated movie," says special effects expert Glenn Derry. Planning great camera shots is the key.

To make your indie shine, Derry advises using a device he hacked together with James Cameron called a "virtual camera." It helps you compose shots in effects movies, and even plan complex shots in traditional live-action films.

The gadget, essentially an LCD screen festooned with motion capture markers, is something you hold while walking around a motion capture studio. It lets you see the animated world of a digital movie or a mockup of a live set.

Derry has created a plugin especially for MAKE readers to make a low-cost virtual camera (note that "low cost" in Hollywood

KEY COMPONENTS

Cintiq 12WX LCD touchpad Wacom (wacom.com), $999

Xbox 360 Wireless Controller Microsoft, $50 list price; available online for $30–$40

V120:Duo motion tracking camera system OptiTrack (optitrack.com), $1,499

MotionBuilder Autodesk (autodesk.com), $3,995

Custom plugin for MotionBuilder download from makeprojects.com/v/27

terms means software and hardware that add up to $7,000; but student discounts and borrowed equipment would bring the price down considerably).

"The indie filmmaker will take longer to finish a film, and will need to bribe his or her buddies in the CG class with beer," says Derry. "But this setup uses the same concept as the stuff we use on *Avatar*." ▨

PEASANT DA VINCIS
Incredible inventions from Chinese villagers.
By Tom Vanderbilt

In 2004, the Chinese artist and collector Cai Guo-Qiang began to hear stories about fantastic "peasant inventions" trickling out of the Chinese countryside: submarines, airplanes, robots, and even "UFOs," fashioned in tiny backyard workshops by people without easy access to technical information, tools, and proper materials.

The following year, Cai would acquire his first invention, a fish-shaped submarine — replete with eyes, fins, and painted depth markings on the side — named *Twilight No. 1*.

The sub, which is pedal-power-propelled by its pilot, is one of a number of submersibles built by Li Yuming, a 70-year-old with a grade-school education who lives in Wuhan. Having mortgaged his house to fund his work, Li rises each day to tinker with his submarines,

working just outside the door from his wife, who, as Cai recounts, "watches his shadow at work."

The life of a would-be peasant inventor is not an easy one, Cai tells me over breakfast at The Peninsula hotel in Shanghai one morning. When Li built a second, larger submarine — named, naturally, *Twilight No. 2* — and set out to sail it down the river, "the government was worried that he would have a liability issue if it sunk," Cai says. Authorities were also concerned it would clog the traffic of local waterways. "So the government river authorities towed the boat to a subsidiary river where there's less traffic, and just let it sink."

As an artist, Cai has become known primarily for his work with the traditional Chinese media of gunpowder and fireworks (I first met

Justin Jin (*Twilight No. 6*), courtesy Cai Studio

GIZMOS FROM THE PROVINCES

⤢ (Opposite) Du Wenda's *Flying Saucer D*, fashioned from aluminum, propeller blades, and LEDs, on the rooftop of the Rockbund Art Museum in Shanghai.

⤢ (Left) Nine inventors with the painted slogan *Peasants — Making a Better City, a Better Life*, by Cai Guo-Qiang.

⤢ (Below) Li Yuming's *Twilight No. 6* is suspended as part of the installation *Fairytale* at the Rockbund Art Musuem.

him a number of years ago at the Fireworks by Grucci test compound on Long Island, N.Y.); his largest audience came as organizer of the fireworks program at the Beijing Olympics.

But he also has something of a curatorial and collecting bent (among other things, the Socialist Realist work of Konstantin Maksimov, a Russian artist who toured China in a moment of Sino-Soviet outreach).

After Li's submarines, Cai went on to purchase any number of Chinese mechanical readymades, ranging across the country-side to distant villages, buying functioning airplanes in Sichuan province (the inventor, Wang Qiang, used plastic bathroom drainage tubes for the fuel tank), rickshaw-pulling robots in Tongxian, flying saucers in Xiaoxian, and even a working submarine made from fused oil drums in Fuyang.

The bulk of these devices were displayed this past summer in the show *Peasant Da Vincis*, at the Rockbund Art Museum in Shanghai, set amidst propagandistic banners that proclaimed aphorisms inspired by the

creators themselves, e.g., "Never learned how to land."

Atop the building spun a (non-)flying saucer by Du Wenda, while on the second floor, Wu Yulu conducted a "robot workshop" demonstrating a delightful range of automata, from a chess-playing contraption to a robotic interpretation of artist Yves Klein's famous *Leap into the Void*. A somber note was struck by the placement of a shattered motor, the remnants of a plane built by Tan Chengnian, who died in 2007 after he crashed his third airplane.

"There's something common shared between all these inventors," Cai says. "They want to fight their gravity, and the restraint of the circumstances that they find themselves in." The act of creation is more important than the result. ("What's important isn't whether you can fly," announced one banner.)

While five of the inventors who came to the show's opening had created airplanes or other flying machines, Cai notes that for three of them, the trip to Shanghai was their first

commercial flight. When Cai asked Wu Shuzai, the creator of a rough-hewn wooden helicopter (which some compared to a chicken coop) with rotor blades repurposed from a threshing machine, where he wanted to go if he could get his craft aloft, his answer was the capital of Jiangxi province — basically a few towns over. "That's where he wants to go," Cai says. "The nearest idea of a city he has is a country town."

The notion of DIY home industry in China invariably recalls Mao Zedong's doomed drive, during the Great Leap Forward, to ramp up steel production through a 600,000-strong network of "backyard furnaces." What the program had in revolutionary fervor it lacked in logistical organization and smelting know-how — the resulting iron output was substandard and essentially had to be scrapped.

Where *Peasant Da Vincis* diverges from this precedent — and from the enormous Chinese manufacturing power also on display in Shanghai last summer at the World Expo — is that with these inventions, "peasants are

DREAMS TAKE FLIGHT

✐ (Top, left) Wu Yulu's 13"-high *Walking Robot*, made of repurposed electronics and materials, stands at attention.

✐ (Right) Wu Shuzei built his *Wooden Helicopter* with salvaged lumber, polyester tarp, and a gas engine. His wife dismantled his first copter for firewood.

✐ (Bottom) *What's Important Isn't Whether You Can Fly*, proclaims this painted calligraphy by Cai Guo-Qiang.

trying to find their own voices and their own creativity, courageously in a very controlled environment."

China, says Cai, "is desperately trying to transition from a society where everything's 'made in China' to where things are 'created in China.'" Among these backyard boffins may be some future Nobel Prize winner.

But in the end, Cai's desire is more immediately personal. "What I'm really collecting is my childhood dreams."

Tom Vanderbilt is the author of *Traffic: Why We Drive the Way We Do* and *Survival City: Adventures Among Ruins of Atomic America*. He contributes to many publications, ranging from the *New York Times Magazine* to *Wired* to *Artforum*.

Make:
ROBOTS!

One roams around drumming rhythms and loops, another bops to the beat.
A Roomba spy keeps an eye on your house, as a Teleclaw hands you a treat.
Hack any toy to behave like a bot, and meet new friends at Maker Faire.
Welcome, Robot Overlords! Please forgive us if we stare.

» Go to makezine.com/27 to download your free Make: Robots desktop wallpaper.

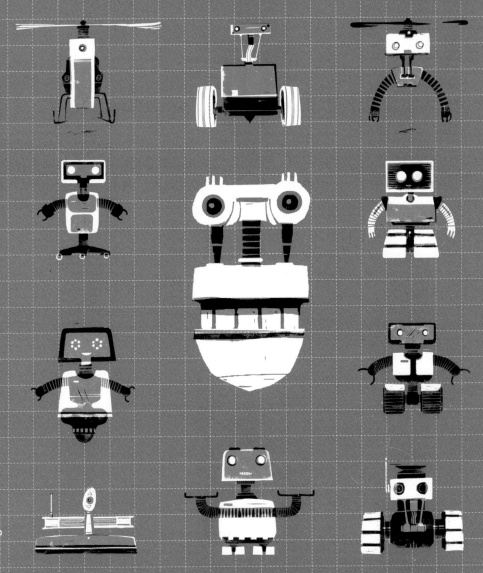

Brian McLaughlin

YELLOW DRUM MACHINE

BY FRITS LYNEBORG

A couple of years ago I started getting into Picaxe microcontrollers, and I ordered a bunch of random components to experiment with. Turning these funny little spring-mounted motors on and off, I thought, Hmm — I could put a stick on that and let the micro-controller drum! This mini sound-recorder board? Hey, if the robot is drumming, it could record itself and play a new beat with the first one in a loop. Oh, nice yellow tracks, a rangefinder — cool, it could move around and look for things to drum on.

After that, the Yellow Drum Machine (YDM) practically made itself. Using hot glue and wooden sticks, it didn't take me long to make a simple chassis, and after a weekend, the finished robot was driving around my house, making wicked beats on stuff it found.

▼
START

1. Build the chassis.
Assemble the Gear Motor and Tread set into 2 tracks with 20 links each. Put a wheel on each motor and solder a 4" (10cm) lead to each motor contact (4 total, Figure A, page 44).

Make a flat wood chassis about 5½"×2¼" (14cm×5.5cm); it should be a bit longer than the tracks. I cut flat sticks into 4 pieces and

glued them together (Figure B). You can also use more sticks, or a single piece of wood with a hole drilled in it for wires to pass through.

Flip the chassis upside down and affix the motors flat to its rear underside with double-sided tape (Figure C). Orient the motors so that the tracks will run parallel along the chassis without touching it or sticking past its rear or front edges. The motors themselves should sit at least ½" away from of the rear edge, to leave room for the GM10 "tail" motor (Step 3). Reinforce the join with 2 applications of hot glue, letting it cool in between.

Prepare your front axle so its wheels will turn freely without touching the chassis. I used 2 washers on each side to space the wheels away from the chassis, and a metal screw hot-glued into each end of the axle to

Garry McLeod

BUILD A FUNKY LITTLE
FREE-RANGE DRUMBOT
THAT ROAMS, MAKES
BEATS, AND SAMPLES.

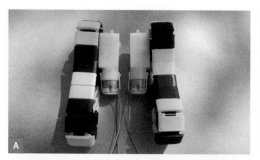

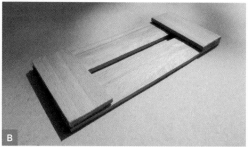

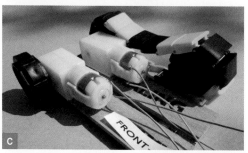

MATERIALS

See makeprojects.com/v/27 for suppliers, prices, and other sourcing information.

Picaxe-28X1 Starter Pack (USB) includes microcontroller in a project board with a programming cable
Motor driver chips, L293D (2)
Solarbotics GM10 pager gearmotors (4) These can only turn from side to side, not rotate fully around.
Sound recorder module (aka sampler board), 20-second such as item #PPM155 from techsupplies.co.uk, or #276-1323 from radioshack.com. The Tech Supplies module's buttons both have wire pairs you can connect to easily. With RadioShack or other modules, you'll have to solder a wire to button pads on the circuit board.
Speaker, 2", 16Ω

Ultrasonic distance sensor (aka rangefinder), 3-pin Devantech SRF05
Solarbotics Gear Motor and Tread Package, GM9 (143:1, 90° shaft)
Flat wooden sticks such as paint stirrers, popsicle sticks, etc.
Lightweight tubing or rod, fiberglass, carbon fiber, or hollow aluminum, about 1' (30cm) for the "bass drum" (tail) and "snare drum" sticks
Lightweight fiberglass rod, very thin, about 6" (15cm) for the "hi-hat" drumstick
Aluminum or other tubing, ¼" (6.4mm) diameter × 4" (10cm) for front wheel axle. Use the same tubing as above or something heavier. I like to cut pieces from old radio antennas.
Metal washers (4) to fit over the axle
Battery holder (long), 4×AA (2×2)
Female-to-female jumper wires (15) Yellow is nice, to match your robot.
Wire, stranded, 18–22 gauge, 3'

Header pins, male, snap-off (4)
Heat-shrink tubing Choose colors to match your design.
Cable ties, various sizes and colors
Rechargeable AA batteries, 1.2V (4) These need to be 1.2V — standard 1.5V alkalines might fry your robot.
Epoxy, quick-setting
Adhesive tape, double-sided either foam or thin

TOOLS

Soldering equipment
Cutting tools for wood and wire
Lighter or heat gun
Screwdriver, small, Phillips for mounting motors
Hot glue gun and glue sticks (5)
Fine-gauge steel wire to hold sticks to the motor shafts while epoxying
Computer Windows, Mac OS X, or Linux

Frits Lyneborg

keep the wheels from falling off (Figure D).

With the chassis still upside down, run the tracks around the wheel pairs. Push the axle forward (away from the motors) as you hot-glue it to the bottom of the chassis (Figure E). Try to get the tracks tight. The chassis will be tilted forward — that's fine (Figure F).

Mount the battery holder to the rear end of the chassis like a rear bumper, using a line of hot glue (Figure G).

Test the sampler board with AA batteries to make sure it works (record your voice and play it back). Cut and strip one end of 2 female jumpers, then unsolder the sampler's mini speaker and replace its 2 original wires with the jumpers (Figure H). Hot-glue the speaker under the chassis at the axle and run the jumpers topside (Figure I).

2. Add the head and neck.

Solder 4 insulated wires, each about 4" (10cm) long, to the back of the SRF05 rangefinder, to all but the fourth pad along the bottom (Figure J). These are the board's 5V, Echo Output, Trigger Input, and Ground contacts.

Epoxy the SRF05 "head" to a GM10 motor, perpendicular across the shaft and pointing away from the small pin sticking up, so the rangefinder can turn from side to side. During curing, I held the board and motor in place with putty and foam tape (Figure K), but you can improvise another way.

Mount the "neck" motor at the front center of the chassis so that the head stands vertical (or leans only slightly forward; not as much as the chassis) and the rangefinder "eyes" sit back 1⁄16" or so (2mm) from the front edge. I used a bit of wood just under 1⁄2" (10mm) to prop the front of the motor to let it sit level, then added liberal amounts of hot glue to secure it to the chassis (Figure L). Try to give the mount a small footprint.

Turn the head from side to side. It should look straight ahead when not touched, and if it moves much more to one side than the other, glue fiberglass rods or wooden stops vertically so that it turns about equally far to the left and right (Figure M, following page).

3. Add the drumsticks.

The robot has 3 drumsticks, each of which is mounted to a GM10 motor: a bass drum "tail" under the battery pack, and thick snare and thin hi-hat drum sticks in front, held by motors attached to wooden uprights just above and behind the rangefinder "head" on either side.

Estimate stick lengths by holding the motors in position and observing which way they'll turn. The bass drum stick (about 2") should hit the ground at an 8 o'clock angle and the 2 front sticks (about 4" each) should be able to reach and drum on vertical surfaces 1" or so in front of the robot. Cut the sticks long, then trim them after seeing how they perform. Shorter sticks work better but look less cool, so you need to find a balance.

Pre-mount each stick to its motor. Drill a hole through thicker sticks to fit over the pin, and tie thin sticks on with wire. Epoxy the sticks in place (Figure N).

Mount the motors to the chassis, eyeballing where you think they'll look good (Figures O and P). For the front motors, I hot-glued wood uprights onto the chassis and added triangular reinforcements in front and back. (No, I'm not sponsored by a hot glue company.) Check that the sticks can travel freely with the head and other sticks in all possible rotational positions (Figure O).

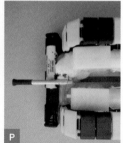

4. Mount the boards and large speaker.

To prepare the Picaxe board, plug one of the L293D motor controller chips into the onboard 16-pin socket, making sure the notch at one end of the chip matches the marking on the board. (Bend its pins to fit the socket by pressing the chip down sideways on a table.) This chip will control the track motors.

Snap off and solder 2×2 male header pins into the 4 holes marked A and B on the board.

It's time to mount the microcontroller, sampler, and larger, cool-looking speaker.

Using more flat sticks and hot glue, I made 2 dividers running fore and aft along the right side to separate the microcontroller and sampler boards and hold them upright on edge (Figure Q). Note that the Picaxe board's

programming jack needs to be easily accessible. On the sampler board, cut off the battery clip, but leave its wires connected to the board.

Then I added a pillar on the left side and a roof to enclose the boards and support the large speaker on top, pointing up.

Rangefinder SRF05

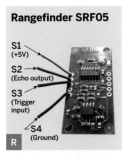

S1 (+5V)
S2 (Echo output)
S3 (Trigger input)
S4 (Ground)

R

External motor controller

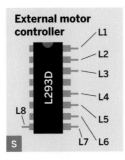

L293D

L1
L2
L3
L4
L5
L6
L7
L8

S

Microcontroller board AXE20

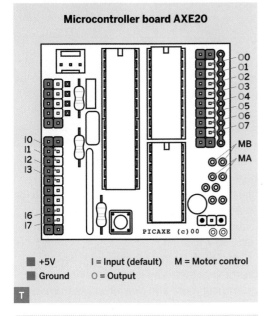

00
01
02
03
04
05
06
07

MB
MA

I0
I1
I2
I3

I6
I7

PICAXE (c)00

■ +5V
■ Ground
I = Input (default)
O = Output
M = Motor control

T

cutting female-to-female jumpers in half and soldering each half to a plain wire if more length is needed. (You can also connect ground wires to the AXE20 board via the solderable pads along one edge.)

» Connect the motors (tracks, drums, and neck) either way for now. You'll run test routines later and swap the 2 connections to any motors that run backward.

CONNECTIONS

Right and left track motors: Connect to the onboard motor controller — left motor wires to the 2 MA pins on the main board, and right motor wires to the 2 MB pins. You'll swap pins later if motor direction is reversed.

Rangefinder: S1 to +5V, S4 to Ground, S2 (Echo) to microcontroller I1, S3 (Pulse) to I0. The code converts pin I0 to an output, so it can trigger the pulse.

External motor controller: L1, L7, and L8 to +5V, L4 to Ground, L2 to I6, L6 to I7. The code converts pins I6 and I7 into outputs.

Neck motor: To external motor controller L3 and L5; swap them later if they're reversed.

Drum motors: Snare drum motor to O1 and Ground, hi-hat motor to O2 and Ground, bass drum motor to O3 and Ground.

Small speaker (under chassis): O0, Ground.

5. Wire the rangefinder, motors, and microcontroller.

Now let's wire everything up. Figures R–T show the wiring points for the rangefinder, microcontroller board, and external motor controller chip (for the neck motor). Connect these components with wires, following the Connections list. (Save the sampler board for later, because you need to test it first to determine how it should connect.)

» You can connect voltage wires (V) to any +5V point, and ground wires (G) to any Ground point, whatever is closest or easiest.

» Connect wires to the rangefinder and external chip by soldering and insulating with heat-shrink tubing. Make connections to the AXE20 board by plugging in female header wires. Make your own female headers by

6. Wire the sampler board.

You already cut off the sampler board's battery clip and speaker, but it still has a microphone, an LED, and Record and Play buttons connected by wire pairs. Mount female headers on the black and red wires that connected to the battery clip, then plug red onto a +5V pin on the main board and black onto a Ground. Mount the larger speaker on top of the robot, pointing up, and solder the 2 sampler board speaker wires to it.

Insert batteries in your robot and test that the Record and Play buttons still work. Cut the Play wires close to the button and touch one of them to Ground on the main board. Does it play back? If so, this is the wire you'll need, and you can trim off the other one close to the board. If not, try the other. (With some samplers, you'll remove the Play button, test its pads on the circuit board, then solder a wire to the pad that works when connected to Ground.) Do the same with the Record button — find the wire that works when touched to Ground, and cut the other one away.

Mount female headers to both wires, then connect the Record wire to I2 on the main board and the Play wire to I3.

Mount the microphone and LED somewhere nice and visible. Now's the time to be creative with extra little decorative parts. (I found a shiny plastic ring that works well holding the microphone.) To tidy up, use cable ties and hot glue, and you can also shorten wire lengths.

7. Program, test, adjust, and play.

At this point, you should have a nice little robot. Now it's time to give the robot its program. Download and install the Picaxe Programming Editor or AXEpad (depending on your computer OS) from rev-ed.co.uk/picaxe/software.htm, if you haven't already.

Launch the software, then open the Mode and Ports tabs under the Options panel to specify the type of chip (28X1) and the COM port for your computer's USB.

Switch the main board off (remove a battery) and connect it to your computer using the Picaxe programming cable. Download the code files from makeprojects.com/v/27.

Each motor has a test code file in the test directory. Open each in the programming editor and click Program to compile and upload it to the microcontroller. Replace the battery and check that the motor behaves as it should, as described at the top of the test code, or swap its wires if it runs in the reverse direction.

The test runs will also guide you in trimming the sticks to adjust their balance. The sticks should be light enough for the motors to move easily, but with fiberglass or carbon fiber rods, it might help to epoxy a small bolt or screw at the end, to make the strikes louder.

With all the motors tested and wired properly, compile and upload the final code, ydm_default.bas. Place the robot on a hard surface and power it up; it should look to the sides and

begin driving around. After some time, it may feel it's time to get funky. It will then search for an appropriate item (like a wall) to drum on. Eventually, it will find an object it likes, then tap a little beat on it.

The Yellow Drum Machine will record its drumming, then play it back in a loop while drumming more and making clicks from its lower speaker to the beat. Meanwhile, it also makes dance motions with its head and body.

While the robot records itself, it will also record your voice or other nearby sounds and include them in its tune. This is especially entertaining for kids. Play a musical instrument to "jam" with the YDM, letting it record and play back your riffs as an accompanist.

The BASIC software is written for modifying and playing. All parameters can be reset, and notes at the top and comments throughout explain how everything works. So you can make the YDM drum only when you enter the room, or use it as a platform to do a lot of other things. You can even make it draw.

Frits Lyneborg is co-host of "The Latest In Hobby Robotics" (blog.makezine.com/video) and runs letsmakerobots.com, the largest online community of its kind, which he started in 2008 as a forum for robot electronics, programming, and funny ideas and inspiration.

ROOMBA RECON

BY RAYMOND CARUSO WITH EVIN PAPOWITZ

TURN AN OLD ROOMBA
INTO A WEB-CONTROLLED
WIRELESS REMOTE
SURVEILLANCE VEHICLE.

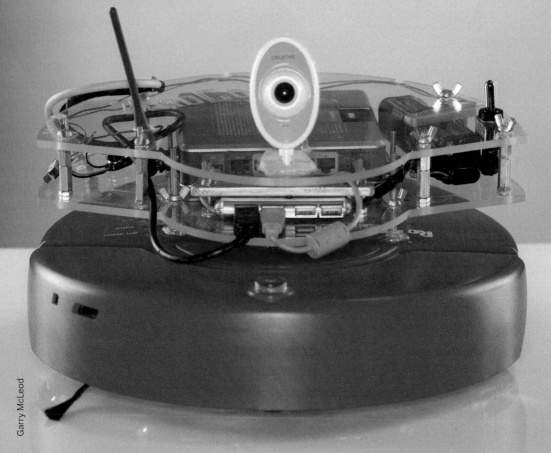

Garry McLeod

Who said a Roomba was just for vacuuming? For robot makers, it's an excellent platform for modifications. Why not start with an already-working autonomous vehicle? We collaborated with our friend Rob Schumaker and decided to turn an old Roomba into a remote surveillance vehicle, which Raymond now uses to spy on and play with his dog while he's at work.

The key to the Roomba Recon's operation is that it carries a wireless router on its back. The router acts as the Roomba's "brain." Its wi-fi side communicates with a nearby computer, which sends it instructions and receives captured images. Meanwhile, the USB side of the router connects to both the Roomba itself and a USB webcam "eye" on top through a small USB hub. The router gets power from its own onboard battery pack.

To enable the router to interface with the computer, Roomba, and webcam, we first replaced its native firmware with the open source OpenWrt. This allowed us to install 2 more pieces of free software on the router: RoombaCMD, which enables control of the robot through a serial port on its side (available at roombahacking.com/roombahacks); and Spca5xx, a device driver for the webcam from the folks at mxhaard.free.fr. The webcam plugs into the USB splitter directly, while the Roomba connects via a hacked USB-TTL cable outfitted with an 8-pin mini-DIN plug that fits the Roomba's serial port.

MATERIALS

See makeprojects.com/v/27 for recommended suppliers, prices, and other sourcing information.

Wireless router, Linksys WRTSL54GS You could use another router compatible with OpenWrt (see wiki.openwrt.org/toh/start for a list), but it might need different drivers.
iRobot Roomba robotic vacuum cleaner We used an old Roomba Red 4100 model (Discovery/400 series).
Webcam, compatible with Spca5xx driver See mxhaard.free.fr/spca5xx.html for a list. We used a cheap Creative Live Cam.
USB hub, 4-port, aka splitter
TTL to USB serial cable, TTL-232R We used #TTL-232R-5V from FTDI (ftdichip.com).
8-pin mini-DIN cable
DC power plug, size M coaxial aka barrel connector
9V battery snap connector
Battery holder, 8xAA
AA batteries, 2,700mAh (8)
Diode, 1N4001
Switch, SPDT (single-pole double-throw) toggle
Electrical tape
Windows computer with wi-fi and internet connection Other platforms should also work, using their own Telnet/SSH and FTP utilities.

For mean, green LED/acrylic bling (optional):
Acrylic/plexiglass sheet, clear, 3⁄16" thick, 36"×30"
9V battery snap connectors (2) in addition to 1 above
LEDs, green (4)

Resistor, 330Ω
Switch, SPST (single-pole single-throw)
Machine screws, #10-24, 5" (16) These can be cut down to size as necessary.
Wing nuts, #10-24 (14)
Rod couplings, ¼-20×7⁄8" (14)
Machine screw nuts, 3⁄16" (19)
Flat bar, about ½" wide (2 pieces) to secure the USB hub and battery holder. We cut aluminum plate to size, but anything will work — wood, plexiglass, etc. We secured the hub with a bar, 2 machine screws, and wing nuts; and the battery holder with a bar and 2 inverted machine screws, no wing nuts.
Hot glue (optional)

TOOLS

OSMO/Hacker serial interface (optional) Needed for older Roombas; see makeprojects.com/v/27 for details.
CAT5 network cable aka Ethernet cable
Soldering iron and solder
Multimeter
Wire strippers
Drill and drill bits for machine screws
Screwdrivers to match screw type
Cutter or saw for acrylic (optional) We used a homemade CNC cutter.

Finally, every project needs LEDs. So we added a few to make this thing light up the darkest of rooms. We mounted 4 green LEDs around the back of the custom-cut clear acrylic frame that carries the project components. Engraved on top of the plexiglass is our engineering team name, Zero Cool, from that awesome movie *Hackers*.

▼
START
1. Make the USB-to-serial cable.
To connect the Roomba to the router, we constructed a USB-to-serial tether by splicing together 2 cables: a USB-TTL converter cable with an FTDI FT232R serial chip in its female USB end, and an 8-pin mini-DIN cable that fits into the Roomba's serial jack.

With the TTL-232R cable, cut off the serial end (with a straight 6-pin header) and strip the black, orange, yellow, and green wires (brown and red are cut short). With the 8-pin mini DIN cable, cut off the female end, peel away some of its outer sheathing, and strip the green, yellow, white, red, brown, and orange wires (cut the other 2 short).

Use a multimeter set to "continuity" to trace which pins on the mini-DIN plug go to what color wire (Figure A). Write it all down.

Solder the wires from the mini-DIN cable to the TTL-232R according to Figure B. The orange (transmit data, TXD) and yellow (receive data, RXD) wires from the TTL-232R cable connect to pins 5 and 3 of the mini-DIN plug, respectively. Black (ground) connects to pins 7 and 8, green (device detect, DD) to pin 6, and V+ to pins 1 and 2.

If you want to steal power from the Roomba battery for lighting the LEDs, as we did, solder a 9V battery snap between V+ and ground (Figure C).

2. Enable the Roomba serial interface (if necessary).
On newer Roombas, the serial interface (the Roomba Open Interface, or ROI) is enabled out of the box. But for models manufactured before October 2005, like ours, you can only

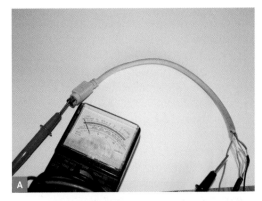

A

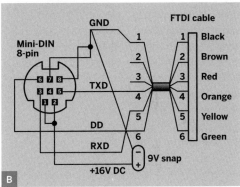

B

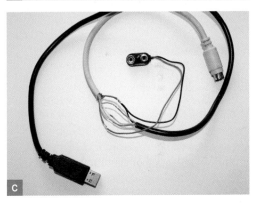

C

communicate through the robot's serial port after flashing its firmware with an OSMO/hacker device.

To use the OSMO/hacker, remove the serial port cover on the Roomba's side. Plug the OSMO/hacker into the port and watch its LED blink for about 90 seconds. Then unplug the device after the light stops blinking.

3. Install OpenWrt on the router.
Our router "brain" will drive the Roomba,

Raymond Caruso

but first we have to open it up with the third-party software, OpenWrt. When installed as a router's firmware, OpenWrt not only lets the router create wireless networks or join existing ones, but enables it to run additional software like RoombaCMD. Furthermore, OpenWrt has a built-in Linux web server that serves pages at the router's own IP address (http://192.168.1.1). This lets us use a browser to both configure the router software and, later, control the Roomba remotely.

To install OpenWrt, visit downloads. openwrt.org/whiterussian/0.9 and download *whiterussian-0.9.tar.bz2*. There are several builds available, but White Russian was the most stable with the RoombaCMD code and our device drivers. RoombaCMD won't run on the more recent Kamikaze builds unless it's recompiled with new libraries.

Connect your computer to the router with an Ethernet cable, then point a browser to http://192.168.1.1 to access the router settings. To log in, leave the username blank and use the password `admin`. Then click Administration → Firmware Upgrade.

Click Browse, navigate to the firmware you just downloaded, and click Upgrade. Don't turn the router off; it will reset itself when complete, after at least 5 minutes.

Reload http://192.168.1.1 in the browser to open the OpenWrt dashboard and set the router password. If the password can't be set on the web interface, open a command prompt or use a Telnet client to telnet to 192.168.1.1, then enter `passwd`.

Click the Wireless tab and either specify Ad-Hoc Mode (Figure D) or Client Mode, select your existing home wireless network for ESSID (extended service set identifier), and fill in your network settings. This setup will let our computer talk to the router, which we'll be doing a lot of.

4. Make the battery pack.
To avoid our router's having to use up precious Roomba battery power, we decided to run it off its own battery pack. Our Linksys WRTSL54GS router used a 12V DC wall wart

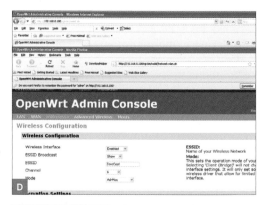

OpenWrt Admin Console

D

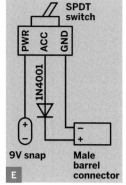

E

F

adapter, so we substituted an 8xAA battery pack and added an SPST switch to make the process of restarting the router a little easier.

Solder the cathode side (marked with a stripe) of a 1N4001 diode to the positive side of a DC power plug that fits your router (Figure E). This will prevent reverse voltage from frying the router. Solder a short wire between the diode's anode and the middle/common pin (sometimes marked "ACC") of an SPDT switch.

Solder a wire between one side of the switch and the red (+) wire of a 9V battery snap. Solder wires to connect the opposite side of the switch to the snap's black (−) wire and the negative side of the barrel connector. Insulate all connections with electrical tape (Figure F). The snap can now be attached to the battery pack.

5. Install the USB and serial drivers.
Now it's time to start installing drivers on the router. To do this from our Windows laptop,

we used the free SSH client PuTTY, which you can download from the.earth.li/~sgtatham/putty/0.60/x86 as *putty.exe*.

The first drivers to install, *kmod-usb-serial* and *kmod-usb-ohci*, and the FTDI driver enable the router to control its USB port and the FTDI chip on our hacked cable.

Start by connecting the router to your computer with an Ethernet cable to check that your computer has internet connectivity. Launch PuTTY and start a session connected to IP address 192.168.1.1 via your router, using the password you set earlier (Figure G).

In the PuTTY session window (Figure H), enter the sequence of commands listed at makeprojects.com/v/27 to install the drivers using the *ipkg* utility. OpenWrt knows where to download most of its drivers, so when you enter ipkg install [package-name], it looks up the package's location and automatically downloads it from there.

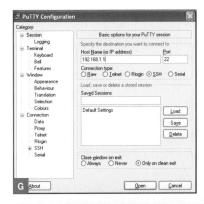

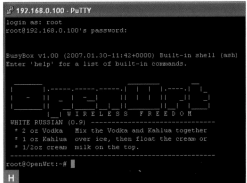

6. Add the webcam eye.
A USB webcam allows the Roomba Recon to capture and stream JPEG images using OpenWrt's built-in web server. These images display on a Flash-based page alongside the robot's browser-accessible control panel.

To keep our robot looking sleek, we shortened our webcam's cord (Figure J). To do this, open the camera case and cut the cord about 2" from where it connects to the board, then unsheath the wires. Shorten the cable wires to just a few inches, unsheath, and resolder the wires back together (Figure I). Tape up the solder joints and tuck it all back inside the case.

To install the camera drivers, use PuTTY to connect to the router, as in Step 5, then follow the instructions listed at makeprojects.com/v/27. From the PuTTY session window, you will download and unzip the *spca5xx* webcam driver, then install it and other needed drivers and libraries into OpenWrt's startup folder. You'll also install the *spcacat* utility, which captures the webcam's current image.

Finally, you'll need to install ImageSnap, a simple, three-line script that calls *spcacat* and stores the resulting image file where the

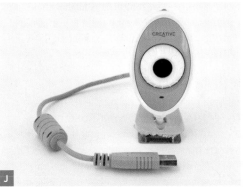

Gregory Hayes (J)

OpenWrt web server can find it. Download *imagesnap* from makeprojects.com/v/27. To upload it to the router, you can use the free file-transfer program WinSCP (winscp.net). Launch WinSCP (Figure K) and connect to your router as you have been with PuTTY. Then navigate to *imagesnap* in your computer's pane and the root (/) directory in your router's pane, and drag-drop *imagesnap* over to the router.

Connect to your router through PuTTY and enter the following commands (type what comes after each # sign, followed by Return):

```
root@OpenWrt:~# chmod +x imagesnap
root@OpenWrt:~# mv imagesnap /usr/bin
root@OpenWrt:~# imagesnap &
root@OpenWrt:~# ln -s /tmp/SpcaPict.jpg /www/
    SpcaPict.jpg
```

This gives the *ImageSnap* script the proper permissions and location, starts it running, and creates a symbolic link so that OpenWrt's web server can find its output.

7. Install RoombaCMD.

RoombaComm, originally written by Tod E. Kurt, is a Java library for communicating and controlling the Roomba from a computer. RoombaCMD, which we'll use, ports this remarkable code to run on a router. It allows us to send commands to the Roomba, making it move as we please. In addition, it has premade templates for the web interface we'll add later.

To install RoombaCMD, connect to the router through PuTTY and enter the following commands:

```
root@OpenWrt:~# wget http://roombahacking.com/
    software/openwrt/roombacmd_1.0-1_mipsel.ipk
root@OpenWrt:~# ipkg install roombacmd_1.0-1_
    mipsel.ipk
```

8. Build the Flash page.

A Flash page published by the router's web server lets you control your Roomba, see through its webcam, and show it off to your friends from anywhere in the world. The Flash interface is encoded in 4 files, which can be downloaded from roombahacking.com/

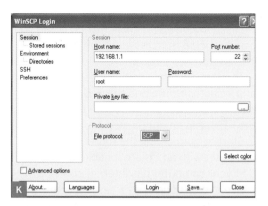

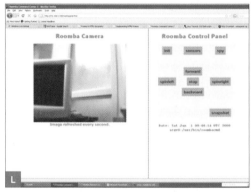

NOTE: You need to run imagesnap & **each time after you reboot the router (or you can add to the startup script; see** makeprojects.com/v/27 **for instructions).**

roombahacks/roombacmd/roombapanel. Then download and unzip *RoombaControl.zip* from makeprojects.com/v/27.

Connect to your router with WinSCP as in Step 6 and point the remote pane to the /www directory. Move all *roombapanel* and *Roomba-Control* files to /www, and exit WinSCP.

Now you can control your Roomba by pointing your browser to http://192.168.1.1/roomba.html (Figure L), or use the snazzy Flash interface at http://192.168.1.1/roombapanel.html (Figure M).

9. Take it worldwide.

The next step is to control the Roomba from anywhere with internet access. To do this, connect the Roomba router to your home wireless network and open up a port on your home router to redirect any traffic coming in from port 80 to the Roomba router.

With this working, you can access the robot's web page under your WAN address (the IP address that your ISP gives you) rather than at http://192.168.1.1. See makeprojects.com/v/27 for more detailed instructions on how to forward your router ports.

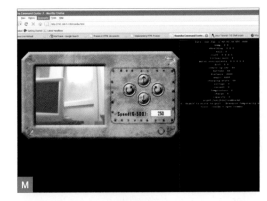

10. Make it mean (optional).

You can piggyback the router, camera, and USB splitter hardware on the Roomba any way you want, but we decided to add some bling by mounting them inside a Roomba-shaped plexiglass frame trimmed with green, glowing LEDs around the back (Figure N).

We used a homemade CNC to cut 2 acrylic panes and engrave the top one with our team name, Zero Cool (Figure O). Then we joined the panes using bolts, wing nuts, and rod couplings to sandwich the electronics in between.

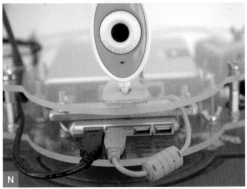

The plexiglass assembly is attached to the Roomba by 4 inverted machine screws that run through the bottom piece of acrylic and are hot-glued to the top of the Roomba. There are other ways to attach it, so be creative!

For the LEDs, we used ledcalc.com to determine that the current-limiting resistor needed for four 2V/20mA LEDs connected in series with a 14.4V supply voltage was 330Ω. The LEDs are connected together with hookup wire to a 9V battery snap that easily connects and disconnects with the 9V snap wired to the router's battery pack and switch.

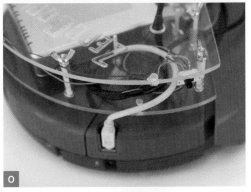

Now what trouble can you cause with your new remote reconnaissance vehicle? The possibilities are endless. Every robot project is improved by some sort of weapon, so perhaps future versions of the Roomba Recon could incorporate something like Luke Cole's USB missile launcher?

✚ Visit makeprojects.com/v/27 for further instructions, resources, and the aforementioned USB missile launcher.

Raymond Caruso is an IT professional from Connecticut. A student at Iona College in New York, he enjoys all types of projects, particularly those that modify existing products.

Garry McLeod (P)

SPAZZI: A SOLENOID-POWERED DANCEBOT

BY MAREK MICHALOWSKI

At BeatBots, we believe that dancing is one of the most worthwhile occupations a robot can have. We like making cute characters move in interesting ways, and as you'll see, this doesn't require expensive components or complicated programming. For Spazzi here, we decided to forego the rotational motors (servos and steppers) that many robots use. Instead, we went with solenoids, for movement that is fast, linear, and percussive.

We designed our popular teleoperated robot Keepon (pictured in Figure A, page 58) for research and the rigors of a child-filled playroom, so he's made from high-end components and custom-machined metal parts. (A toy version, My Keepon, comes out later this year, and a portion of sales will subidize the distribution of research robots to autism therapy practitioners.)

We designed Spazzi as a bouncy and easy to build robotic character that achieves the same adorable bounciness as Keepon through a simpler mechanism: solenoids and springs controlled by an Arduino microcontroller connected to a computer.

A solenoid is an electromagnet that pulls a rod (or "plunger") inside its coil when current is passed through. Spazzi's physical form is extremely simple; for his lower half, 3 solenoids stand parallel to form a triangle sandwiched between 2 plastic parts — a base and a waist. The plungers run up through the top of the waist and hold up the robot's plastic head, extended by compression springs.

The 3 plastic body parts are based on the Reuleaux triangle — a shape with various interesting properties and uses (search online for more info). I made these parts on a 3D printer, but you could improvise your own body and head parts from any material, including cardboard. The core of the robot is just its assembly of solenoids and springs.

Garry McLeod

BUILD A CUTE ROBOTIC BOBBLE-HEAD THAT DANCES TO YOUR MUSIC, AND EVEN MAKES SOME OF HIS OWN.

MATERIALS

See makeprojects.com/v/27 for recommended suppliers, prices, and other sourcing information.

MAKE Spazzi Electronics Bundle item #MSSPZ from the Maker Shed and Jameco (makershed.com/spazzi), includes:
» Arduino Uno microcontroller, Maker Shed #MKSP4
» MakerShield, Maker Shed #MSMS01
» Mini breadboard, Maker Shed #MKKN1
» Jumper wires, Maker Shed #MKEL1
» Transistors, TIP102 (3) Jameco #32977
» Resistors, 1kΩ (3) Jameco #690865
» Diodes, 1N4004 (3)

Solenoids, sealed continuous pull, 12V, ¾" stroke (3)
Cable ties, 1.8mm wide (6)
Compression spring, type 302, 0.312" OD, 0.02" wire diameter
Latex rubber cord
Electrical tape
Power supply, 12V 2A
3D-printed ABS body parts (3) and eyes (optional, 2)
   3D models for these plastic parts, as well as code files, can be downloaded from the project's page at thingiverse.com/thing:8909. Use a 3D printer to print them out, or send the files to a 3D printing service such as Shapeways, Ponoko, or i.materialise.

TOOLS

Soldering iron and solder
Hobby knife
Diagonal cutters
MakerBot Thing-O-Matic or other 3D printer (optional)
   You can also send the 3D part files out to a service to have them printed for you (see above).

A

B

C

Solenoids are binary: assuming they can draw sufficient power, they are either fully compressed or fully extended. A transistor connected to a digital pin on the Arduino drives each solenoid by supplying 12V from the power supply when its pin is set HIGH. The front solenoid makes the head nod up and down, and the 2 rear solenoids make it lean back to the left or right.

With its 3 solenoids, Spazzi can move to just 8 different positions. But this limited repertoire produces surprisingly rich and varied movement when the activation and frequency of the solenoids are varied over time. This is performed by Max/MSP or Pure Data (PD) software on a computer, which sends on/off commands to the Arduino over a USB/serial port in response to music or other inputs.

Make Spazzi dance with the parts and code listed here, and then after that, you can choreograph his moves however you want!

▼
START

For more step photos, see makeprojects. com/v/27.

1. Assemble the MakerShield according to the instructions at makezine.com/go/makershield. Stick the mini breadboard to the center. Install the MakerShield on the Arduino Uno.

2. Insert the 3 transistors into the breadboard in a row, making sure that they don't touch. Looking at the printed face, the leads from left to right are Base-Collector-Emitter.

Marek Michalowski

D

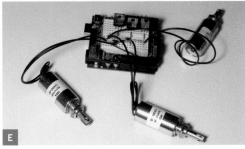

E

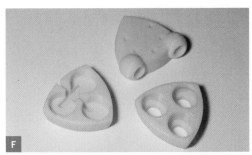

F

G

H

I

3. Using jumper wire, connect the 3 emitter leads to Ground on the Arduino/MakerShield. Connect one end of three 1K resistors to pins D5, D6, and D7, and the other ends to the base lead on a transistor (Figure B).

4. Jumper each transistor's collector lead to its own rail on the other side of the breadboard (Figure C). Insert the anode of each 1N4004 diode in one of the collector rails, and the cathode (marked with a band) into a new rail on the breadboard. Jumper the shared cathode rail to the Arduino's V-in pin and to a new rail on the breadboard (Figure D).

5. Connect one lead from each solenoid into a collector/anode rail. Connect the other leads into the new cathode rail (Figure E).

6. Using a 3D printer, print the 3 body parts: base, waist, and head (Figure F). These took about 45 minutes each on a Thing-O-Matic. If you don't have access to a 3D printer, get creative! You can use found objects, wood, ShapeLock, or even cardboard.

7. Insert the 2 rear solenoids into the base so that their leads run out through the channel. Insert the final solenoid in front, so that its leads run over the 2 rear solenoids' leads and out the back (Figures G and H).

8. Remove the nuts and washers from the solenoids and fit the waist part over them. Reinstall the washers and nuts (Figure I).

9. Remove the plungers from the solenoids and loosely tie a cable tie through the hole of each.

10. Cut three 22mm lengths of compression spring (approximately 6 coils each). With each plunger, thread the pointy end into a spring and rotate the spring until one turn passes under the cable tie at the back end of the plunger. Tighten the cable tie to hold this single turn of the spring snugly against the plunger (Figure J, following page).

11. Feed the cable tie ends through the holes in the head part.

12. Cut the heads off 3 more cable ties and slide them over the cable tie ends in the robot's head (Figure K). Slide them down so there is no slack in the cable ties, but not too tight, and snip off the cable tie ends.

13. Attach the head assembly to the body so that the eyes are opposite the wires (Figure L).

14. Cut six 40mm lengths of rubber cord. I chose this material because it looks similar to the wires in Spazzi's "tail," but you can use anything to make the decorative antennae.

15. Insert 3 lengths of rubber cord into each of the 2 holes atop the robot head (Figure M). If they are too loose, cut a few millimeters of the discarded cable tie and wedge it in with the cord.

16. Using a 3D printer, print the supplied pupil models out of black ABS plastic and affix them to the eye sockets with hot glue. If you don't have black ABS, you can also color the pupils using paint, marker, or circles cut from electrical tape (Figure N).

17. Connect the 12V power supply to the Arduino, and connect the Arduino to a computer with a USB cable.

18. If you're an experienced Arduino programmer, you can probably take it from here — setting pins D5, D6, and D7 to HIGH and LOW will activate and deactivate the solenoids.

Otherwise, upload the provided *Spazzi.pde* patch to the Arduino, then set the baud rate to 38,400 in the Serial Monitor. Entering the letters a, s, and d will activate the solenoids, while q, w, and e, respectively, will deactivate them. You can make Spazzi dance by using the keyboard like this; now you just need an application that sends these characters over the serial port!

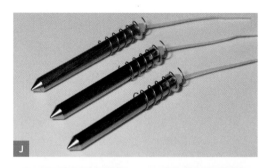

J

K

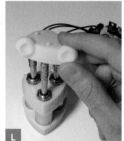

L

M

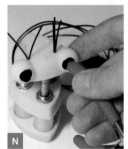

N

Marek Michalowski doesn't shrug off Spazzi's ability to charm people with its dance moves.

Garry McLeod (portrait)

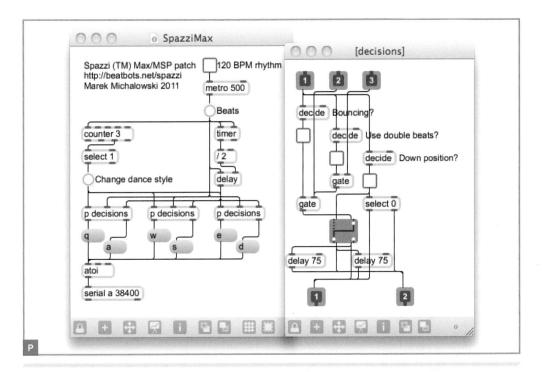

Application

We like using Max/MSP to control our robots. It's a visual programming language in which data flows between objects — switches, mathematical operators, filters, etc. — over virtual patch cables. Download a trial version at cycling74.com or the free variant PD at puredata.info.

Download the demo patcher (the name of a Max/MSP document) *Spazzi.maxpat* at makeprojects.com/v/27 and open it in Max/MSP (Figure P). The serial object at the bottom opens your Arduino's port. Click the 120 BPM rhythm checkbox (or set your own rhythm with the Beats button), and Spazzi will start dancing. His style will change every 4 beats, based on randomly deciding, for each solenoid, whether it is down or up; whether it's bouncing; and whether it makes 1 or 2 bounces per beat. See a video at the URL above.

To make the robot dance to music, you can patch an audio stream through Olivier Pasquet's *op.beatitude~* object (opasquet. fr/dl/op.beatitude~.zip) or Tristan Jehan's *beat~* object (web.media.mit.edu/~tristan/maxmsp.html) and feed the detected beats into your Spazzi patcher. To pipe your music into Max/MSP (or another application) from, say, iTunes, try Soundflower on Mac OS X (cycling74.com/products/soundflower) or Jack on Windows (jackaudio.org). Be creative — Max/MSP and PD are great environments for rolling your own signal-processing apps.

A more advanced challenge is to do the audio processing directly on the Arduino. You'll be constrained by its more limited processing power, but wouldn't it be great to liberate this robot from its tether to the computer?

Next step: Customize! Make Spazzi your own. Use other colors, different antennae and other appendages; even put him on a mobile base! And, most importantly, make videos — who knows, perhaps your dancing robot video will go viral!

Spazzi and its appearance are trademarks of BeatBots, but anyone is free to make the robot for personal or nonprofit use.

Dr. Marek Michalowski (marek.michalowski.me) is a roboticist living and working in San Francisco. He co-founded BeatBots (beatbots.net) with Dr. Hideki Kozima, designer of the robot Keepon.

TELECLAW: REMOTE ROBOT GRIPPER

BY GORDON McCOMB

It might have been *Robots of Saturn* that first got my young brain thinking about building a mechanical man. In that obscure 1962 sci-fi adventure novel, Dig Allen and his fellow teenage space explorers transfer their thoughts into the bodies of teleoperated robots to mine Saturn's dangerous rings for precious Methane-X.

Using some of my stepfather's "extra" ham radio gear, I tried to build my own telerobot, with plans of world domination swimming through my head. I didn't get far, but now, with powerful yet inexpensive microcontrollers, maybe this time it'll work.

Big ideas begin with small steps, so let's start petite, cheap, and easy. Here's how to build a super-affordable remote robotic gripper that can pay the pizza delivery guy.

The telerobotic gripper (let's call it the Teleclaw) has three main parts (Figure A). The mechanics include a $2 plastic clamp, a radio-controlled (R/C) servomotor, some stiff wire, and a bracket to hold it all together. The electronics are a simple circuit designed around the Picaxe 08M microcontroller and an infrared (IR) receiver/demodulator module.

The remote control unit is an ordinary TV/VCR/DVD remote. You can use a remote you already have or get a cheap one just for this project. I paid 99 cents for the model here, but note that cheaper remotes usually have shorter ranges.

▼
START

1. Construct the gripper mechanics.

If you're making your own gripper bracket, use rigid ¼" plywood or expanded PVC, and follow the diagram in Figure B. For a quick prototype you can use ¼" foam board.

Saw out the bracket's basic shape, then drill the mounting holes, all ⅛" unless otherwise noted. For the rectangular cutout (which will fit the servo), you can drill a starter hole and thread a thin saw blade through to cut out the rest, or else just cut through along the dotted lines shown in the diagram.

Secure the servo into the bracket using four 4-40×½" machine screws and nuts, with the shaft end near the bracket's bend (Figure C).

Drill 2 (or possibly 3) ⅛" holes through the clamp handles, one at the end of one grip (my clamp already had this) and 2 more 1" apart along the other, matching the ³⁄₁₆" holes in the bracket (Figure D, page 64).

Use flush cutters to trim the plastic away from the clamp's handle around the red

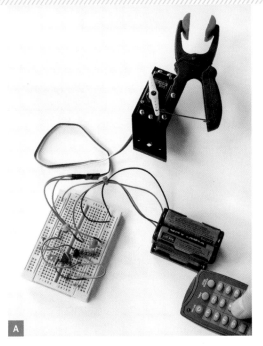

A

USE A TV REMOTE TO GRAB AND RELEASE SMALL OBJECTS FROM AFAR.

MATERIALS

See *makeprojects.com/v/27* for recommended suppliers, prices, and other sourcing information.

MAKE Telerobotic Gripper Kit item #MSTGK from the Maker Shed (makershed.com/teleclaw), includes all of the following project Materials:

Clamp, ratcheting plastic, about 5" long, with jaws that open at least 1¼" such as Sears Craftsman #31594

Servomotor, radio-controlled (R/C), standard size (roughly 40mm×20mm×38mm), with double-arm horn such as GWS S03N STD. Save your money and skip the fancy digital or metal gear servos.

Picaxe 08M microcontroller

Infrared remote control compatible with Sony equipment

Infrared receiver/demodulator, 38kHz Be sure it has a wide voltage range if using the Picaxe at less than 5V DC.

Gripper bracket ready-made or cut from aircraft-grade plywood or expanded rigid PVC, ¼" thick, 3½"×2"

Stiff wire unbent 2¼" coilless safety pin or a 3" length of 18 gauge or thicker steel or brass wire (not annealed)

Solderless breadboard, half-size Maker Shed #MKKN2

Hookup wire, solid core, 22 AWG Maker Shed #MKEL1

Stereo audio jack, 3.5mm

Resistors, ¼W, 2%–5% tolerance: 330Ω (2), 4.7kΩ, 10kΩ, 22kΩ

Capacitors, electrolytic, 15V or greater: 4.7µF, 47µF

LED, T1¾ (5mm) any color

Power supply, 5V DC, regulated, 200mA minimum or battery holder with 3×AAA alkaline batteries (4.5V) or 4×AAA NiCd or NiMH rechargeable batteries (4.8V); for Picaxe.

Power supply or wall adapter, 6V DC, 750mA minimum or battery holder with 4×AA batteries

Machine screws, 4-40 pan head: ½" (4), ¾" (2)

Nuts, 4-40 hex (6)

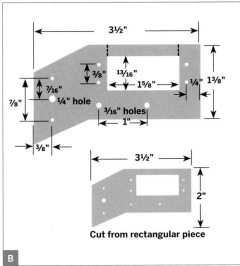

3½"

3/8" 13/16"

7/16" 1 5/8" ¼" 1 3/8"

7/8" ¼" hole

3/16" holes

1"

3/8"

3½"

2"

Cut from rectangular piece

B

TOOLS

Picaxe USB programming cable, AXE027 If you use a different Picaxe cable, you'll need a different connector than the 3.5mm stereo jack specified above.

Saw(s) to cut the bracket; e.g. hacksaw, jigsaw, scroll saw

Drill and drill bits: 1/8", 3/16", ¼" plus a bit to start the cutout for the servo (optional)

Wire cutters (flush cutter type) and wire strippers

Pliers: needlenose and lineman's (heavy-duty)

Computer with internet connection

Soldering iron and solder

Multimeter (optional) to test connections

C

ratcheting piece, then yank out the ratcheting mechanism with heavy-duty pliers.

Mount the clamp to the bracket using two 4-40×¾" bolts and nuts. Make them tight without seriously deforming the plastic.

Drill a ⅛" hole toward the end of one side of a double-arm servo horn and temporarily fit the horn over the servo shaft. Slowly rotate the servo counterclockwise until the motor hits its internal stop, then back off about 5°. Reposition the horn so that its arms are in the 9 o'clock and 3 o'clock positions, then attach it to the servo using the included screw.

Clip off the clasp from a 2¼" coilless safety pin, straighten it, and cut a 3" length (or just start with 3" of stiff wire). Bend ½" of each end of the wire 90° in opposite directions to make a thin "S" hook (Figure E).

Hook the wire through the holes in the servo horn and the outside handle of the clamp. Use needlenose pliers to crimp the wire around the horn and handle loosely, so that it won't bind (Figure F).

2. Build the gripper electronics.

Following the schematics (Figures G and I), wire the circuit together on a solderless breadboard (you can transfer it to a soldered board later). Keep lead lengths short, especially for the 2 capacitors. Leave one side of each battery pack unconnected for now.

For simplicity, this project uses 2 separate voltages, one non-regulated, to operate the Picaxe and the servo. See makeprojects.com/v/27 for how to power the Teleclaw from a single 4.5V–5V supply.

Solder leads to the 3.5mm stereo jack, which the Picaxe programming cable will plug into. Connect the jack and 2 remaining resistors to your board as shown in Figures G and H. All resistors must remain in place even after you've uploaded your program.

3. Set up your remote.

If you're using a universal remote, follow its instructions for how to set it up for a Sony TV, VCR, or DVD player. With my remote, for example, I selected VCR code 098. The Picaxe

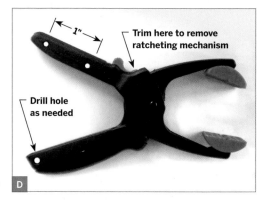

Trim here to remove ratcheting mechanism

1"

Drill hole as needed

D

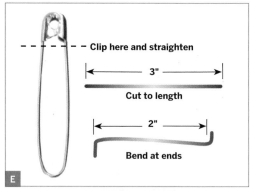

Clip here and straighten

3"

Cut to length

2"

Bend at ends

E

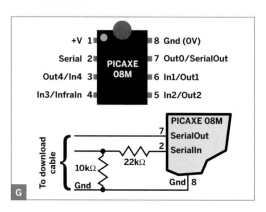

F

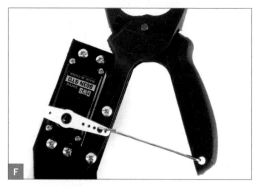

+V 1 — 8 Gnd (0V)
Serial 2 — 7 Out0/SerialOut
Out4/In4 3 — PICAXE 08M — 6 In1/Out1
In3/InfraIn 4 — 5 In2/Out2

PICAXE 08M
7 SerialOut
2 SerialIn
To download cable
10kΩ
22kΩ
Gnd
Gnd 8

G

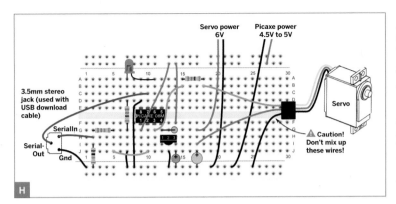

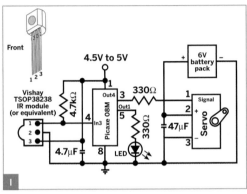

08M has built-in commands for reading and decoding Sony SIRC protocol IR codes.

4. Program the Picaxe.

Download and install the free Picaxe Program Editor (Windows) or AXEpad (Mac/Linux) software from picaxe.co.uk. Plug your programming cable between your computer's USB and the 3.5mm jack.

Launch the Program Editor or AXEpad, and specify 08M under Options → Target Device. Also select the serial port used with the download cable.

Download the project code from make projects.com/v/27 and open it in the editor. Apply power to the microcontroller circuit only — the servo doesn't need to be powered yet — and click the Program button to compile and transfer the code to your Picaxe.

5. Now play!

Connect power to the servo, then disconnect and reconnect power to the Picaxe. The LED should glow to indicate when the Picaxe is sending pulses to the R/C servo, and the servo should quickly center itself.

Point the remote at the IR receiver on the breadboard and start pressing buttons to control the servo actions as follows:

» **Channel Up/Channel Down** — close/open the gripper incrementally.
» **2/8** — close/open the gripper fully.
» **5** — set the gripper to midway.
» **0** — toggle the servomotor power on and off (to extend battery life).

The gripper should be closed, or nearly so, when the servo is rotated all the way clockwise. To adjust its range, unscrew the servo horn from the servo's motor shaft, reposition it, and put the screw back on.

The 0 button toggles "active" mode. Press it once (the LED goes out), and the servo shuts off after each move, saving power and eliminating a slight jitter. Press 0 again (the LED goes on), and the servo receives periodic pulses to set its position, which makes it maintain a tighter grip on things.

Comments in the code explain how to change parameters to fine-tune your Teleclaw.

That's it! It's just that simple. In the next installment of Telerobots for World Domination, I'll tackle the job of creating a weapons system consisting of a pulsed atomic-powered rail gun and chemical laser. Stay tuned, fellow space explorers!

Gordon McComb, "the father of hobby robotics," has been building robots since the 1970s, and wrote the bestselling *Robot Builder's Bonanza*. Read his plans to take over the world with an army of mind-controlled automatons at robotoid.com.

TEACHING OLD TOYS NEW TRICKS

BY DJ SURES

Watching an animated robot is certainly amusing, but interacting with a robot is an experience! You can make interactive robots with unique personalities out of many common toys, and I designed the EZ-B Robot Controller (ez-robot.com) to make the process as easy as possible. This tutorial will introduce how the EZ-B works, and then explain how you can use it to teach an old Digger the Dog pull-toy some new tricks: autonomously chasing a red ball and obeying voice commands.

The EZ-B is a microcontroller circuit board that connects to numerous inputs and outputs. Put one inside a toy, and it will control servomotors, sensors, speakers, LEDs, and other devices that enable your bot's behavior.

Meanwhile, the EZ-B also connects wirelessly to a nearby computer using the Bluetooth protocol. Unlike with other microcontroller boards, all the programming and computational "heavy lifting" with the EZ-B happens on the PC side, and the onboard microcontroller just acts as a slave interface to your robot's motors, sensors, and other peripheral devices. This lets your robot perform voice recognition, speech synthesis, visual feature detection, and other functions far beyond the capability of standard microcontrollers, and it also means you never have to compile or upload new firmware.

Plug-In Peripherals

The EZ-B circuit board has the same general pin arrangement as an ordinary Arduino microcontroller, but instead of connecting the microprocessor chip's I/O pins to single female headers for plugging wires or shield pins into, it breaks them out into 3-pin rows, each with their own voltage and ground.

This lets you solderlessly plug in hobby servomotors with their standard 3-wire female plugs (red = power, black = ground, and white or yellow = signal/data).

Many Arduino-compatible sensors and peripherals also use this type of connector. On the sensor (inputs) side, these include distance sensors, tilt sensors, compasses, thermometers, button pads, and joysticks.

Popular effector components (outputs) with 3-wire plugs include servomotors,

Colin Way

DJ SURES
CONVERTS TOYS
INTO SURPRISINGLY
CAPABLE ROBOTS.

H-bridge adapters for motor control, and high-voltage relay adapters (Figure A).

With the included female-to-female header adapter, the EZ-B is compatible with any Arduino shield. In addition, it has a header for I2C devices, which use 4 wires: power, ground, clock, and signal. This protocol, which is gaining popularity in robotics, lets you chain together multiple low-speed peripherals, ranging from servos to programmable LED modules (like the BlinkM RGB LED), and communicate with each in the software by using its unique address.

The EZ-B also has built-in protocol support for MIDI, iRobot Roomba control, and TellyMate serial-to-TV control.

For powering a robot's wheels or tractor treads, you can often use the toy's original motors and gears. Otherwise, you'll need a *gearmotor* (or DC motor and gear set), or a *continuous-rotation servomotor*. Plain DC motors spin too fast, and the gears decrease motor speed while increasing torque. Continuous-rotation servos are variable-speed forward/reverse motors that use 3-wire servomotor connectors.

For robot joint movements and other animations, (e.g., to control head/camera position), you'll use servomotors. You'll most likely need to replace a toy's original servos, if it has any, with generic hobby modules. Servos make precise movements within a range of 90° to 120°, specified by software. Small servos are fragile, so be careful with them; rotating their shafts manually can strip the gears and render them useless.

Distance sensors are favorite robot input peripherals that can typically detect >1"-wide solid objects from ≅ 2" to 30" away. Ultrasonic sensors, like the Satistronics HC-SR04, are less effective with angled or textured objects, while infrared rangers, like the Sharp GP2D12, are less effective in sunlight, so many robot builders combine both.

On the host computer side, the EZ-B software reads from the computer's USB port in addition to communicating with the EZ-B controller via Bluetooth. This lets you pack

A

MATERIALS

See *makeprojects.com/v/27 for recommended suppliers, prices, and other sourcing information.*

Toy with moving parts
EZ-B .NET Bluetooth Robot Controller V3 $119 from ez-robot.com
Computer running Windows 7 or Vista, with Bluetooth transceiver
Batteries, AA (5)
2.4GHz wireless camera
I got one on eBay for $19 with an embedded battery that charges off USB.
GWS modified servos (2)

Servo horns (2) star-shaped attachments for mounting
Small screws
Optional, for LED and speaker connections:
Servo extension wires (2)
LED

TOOLS

Dremel with cutting wheel
Screwdrivers, small
Sharpie
Masking tape or label maker
Optional, for LED and speaker connections:
Soldering iron and solder

more functionality into your robot by using 2.4GHz wireless USB devices, which don't interfere with Bluetooth. For example, you can install a wireless camera onboard your bot and plug its wireless dongle into the host computer to have your EZ-B program process the live video feed. By using a 2.4GHz wireless USB hub, you can support both a camera and a sound card (to give your bot better sound), along with any other USB peripherals.

The EZ-B has an SDK (software development kit; ez-robot.com/sdk) that supports development for the controller within the larger EZ Robot Project. Community contributors have developed EZ-B software modules for a growing number of peripherals and sensors, letting you control these devices from the EZ-B without worrying about the low-level electrical details of the connections.

DJ Sures

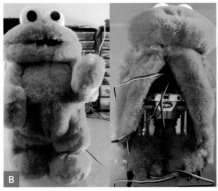

B

▼ START

1. Set up your environment.

With any robotics project, you'll want to focus on the creative side. So before you begin building your robot, create a quiet, comfortable, and clutter-free workspace where your ideas can flow.

It's discouraging having to stop midway because you're missing some supplies, so make sure you have enough hot glue sticks and a variety of screws, nuts, and bolts around. (Keep leftover screws from any toys you modify, along with any reusable lights, motors, gear sets, springs, and switches.)

If you'll be Dremeling hard plastic, be prepared for a mess of shavings, and make your cuts outside or in a room with easy vacuum access and no carpet. If you'll be working on a table, cover it with cardboard first. This prevents scratches and messes from the hot glue gun, and is also handy for making quick sketches or notes.

2. Find a toy.

Browse thrift stores or revisit your old toy box, and find a toy that has moving parts and good robot-control potential. These can be robots, vehicles, animals, or even dolls (Figure B). I should warn you, however, that converting a doll into an animated robot may result in disturbing some of your friends.

The toy that leaped out at me when I was hunting for ideas for this article was Digger the Dog, a $7 pull-toy from Playskool. Here's how you can get Digger to chase a red ball and obey voice commands, and you won't even need to solder or write any code. The toy doesn't even have a motor, so it's also a great example of how to add autonomous mobility.

Take a good look at the toy you'll be modifying. Study its limbs, head, and any other movable parts, and decide what you want to control. Animating arms, head, and other parts that require precision positioning will require servos. As a beginner, start small by adding basic movements. You'll be able to add more features as your robotics experience grows. Practice makes robots!

The most common mechanisms for hobby robot mobility are wheels or tracks. Making a robot that walks is too complicated to explain in this tutorial, unless you start with a toy that already walks via a wheel-driven shuffle motion. In that case, it's easy. To make a stuffed animal walk, you can wrap its "skin" over the shell of a modified walking robot toy.

If you already have your motors and other components, hold them against the toy in order to help you visualize where you'll need to modify it and how to mount everything. With hobby servos, it helps to snap photos of available mounting brackets and motor shaft attachments at the hobby shop, and then refer to them while browsing toys. This minimizes buying hardware that you don't end up using.

3. Add the servos.

Most toys use small Phillips screws, so make sure you have some small screwdrivers. Keep each type of screw you remove in a different small container to avoid confusion when reassembling. Digger had 6 screws under its belly, which lifted off to expose an open area perfect for housing the robot controller and battery (Figures C and D).

The toy may be a little dirty if you purchased it from a thrift store; if so, you can put the plastic parts in the dishwasher once they're dismantled. Remove the pull-rope leash connected to Digger's collar; you won't need it anymore.

Hold the servos and sensors to the desired mounting positions and sketch with a Sharpie marker any cuts or drill holes before you start modifying the toy shell. Follow the carpenter's rule: *Measure twice, cut once!*

With Digger, the servos could fit back-to-back in the rear axle area, provided that the rear wheels were pushed out a bit wider than they sat in the original toy. So I removed Digger's rear axle, dug plastic out of its body underside to make room for the servos, and dug out the rear wheels to accommodate the servo horns (Figure E). I also trimmed back the nose to expose the camera lens.

I secured the servo horns to the wheels with small screws saved from previous toy builds, then hot-glued the servos in place (Figure F). To avoid lopsided wheels, double-check their alignment with your servos.

For mounting nearly all peripherals, a hot glue gun will be your best friend. Fit and hold the servos and sensors against the modified area of the body. Does it look OK? Great! Use a few drops of glue at first, verify the fit, and then go wild and add glue from all available angles.

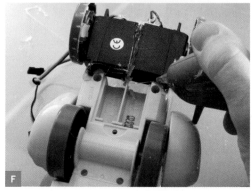

◹ TIP: If you make a mistake, wait for the glue to dry and use a drop or two of rubbing alcohol to separate.

4. Add the speaker, LED, and camera.

For any peripheral devices that don't have 3-pin connectors, including any LEDs, speakers, and other devices that you liberate from the toy itself, I like to add servo connectors to make them easily pluggable. To do this, cut the male end off a servo extension cable and solder to your peripheral, referring to its datasheet to find its power, ground, and signal contacts. For LEDs, speakers, and other 2-wire devices, connect just the signal and ground.

For any other LEDs, springs, washers, mounting hardware, or screws that you liberate, this is a good time to start your collection of miscellaneous hardware.

If you want to connect the speaker, unscrew it, then unsolder its 2 wires and replace them with a servo extension cable, soldering its white/yellow and black wires to each contact (Figure G).

For the LED, wire another servo cable to the LED, black (ground) side to its cathode, which is indicated by the shorter leg and flat part of the plastic. Glue the LED into the hole in the top of Digger's collar, where its leash used to connect (Figure H).

To help you during assembly and in the future, label the cables and plugs of all your peripherals using a label maker, or just masking tape and a marker. Label each servo by its position (right/left) and each sensor by its type. For Digger, we have a left servo, right servo, speaker, and LED.

For the camera, unscrew its case and remove its internal workings. Doing this with any peripheral device saves room, making it easier to fit. Use hot glue to lightly secure the camera inside the toy shell, with the lens pointing out of the nose hole (Figure I). Don't use too much glue, as you may want to realign the camera or reuse it later in a different project.

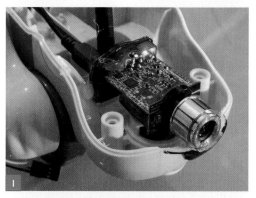
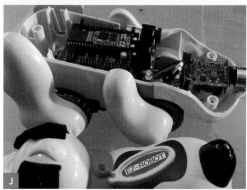

5. Final assembly.

Determine the mounting position of the robot controller and battery pack. With Digger, its belly area was the perfect width to host the EZ-B Controller with the battery underneath. Refer to the EZ-B manual to determine the connection points for each servo and sensor.

Digger's left servo plug is connected to pin row D14 and the right servo is connected to pin row D13. The speaker signal and ground are connected to D7. The LED anode is connected to the signal wire and the cathode to ground on pin row D6. Keep track of all of these connections; you'll need to know them when configuring the software.

Once the connections are completed, carefully reassemble the shell, trying to protect the controller connections and any other fragile parts (Figure J).

To give your robot a unique look, you can paint the shell. Scuff the smooth finish with sandpaper and then clean it to allow the paint

to bond. I painted over the Playskool doggy tag with red paint, then replaced it with an EZ-Robot label from our trusty label maker. You can also paint the entire shell a custom color. Bring it to a paint store and ask them what type of paint and process they suggest.

6. Connect to the PC.

Right-click the Bluetooth icon that should appear in the System Tray and select Add Device to bring up a listing of nearby

Bluetooth devices. Power on the EZ-B, wait for the PC to discover it, then select the device and press Next. When prompted for a pairing key, select Provide A Key and enter 1234. Your computer should add the EZ-B connections as 2 COM ports. Note the lower numbered port; this port will carry the data connection between the PC and the EZ-B.

Download and launch the EZ-Builder software from ez-robot.com/ez-builder. This is the application through which you can program your robot (you can also use Visual Studio .NET, msdn.microsoft.com/en-us/vstudio). Under the Add Control menu, choose Connection, specify the COM port just noted when you established the Bluetooth connection, then click Connect. Then select Servo (also under Add Control) to specify the left and right servo connections at pins D14 and D13, respectively.

Time for our first test. From the Windows menu, add a Modified Servo Panel to your workspace. Press Config and select the Left and Right modified servo ports, then press Save. If all's well, you should now be able to drive Digger around by clicking the arrow buttons or using the arrow keys on the keyboard.

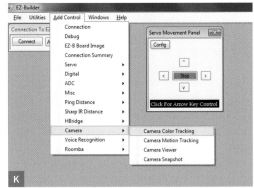

7. Configure ball-following behavior.
If your mobility was a success, install your camera's driver on the PC. Click Add Control → Camera → Camera Color Tracking (Figure K). Your camera should appear in the Video Device dropdown list. Select your camera, then put the ball in front of the lens. Select the ball's color from the next dropdown box (I picked red) and adjust the Color Brightness slider at bottom right until you can only see the ball in the preview window. (A bright room or sunlight works best for color camera tracking.)

To translate the color tracking into motion tracking with the servos, go back to the Servo Movement Panel, click the Config button, and enable Motion Tracking. Then return to Camera Color Tracking and check the Enable Forward Movement box. This will make the robot move forward when the detected colored object is directly in front.

Now you have a robot dog that follows a red ball (Figure M). You'll almost certainly be entertained for a while, but what if the ball is out of view? Instead of having to steer Digger around manually, you can get it to respond to voice commands.

8. Configure voice recognition.
EZ-Builder lets you access the Voice Recognition functionality that's built into Windows 7 or Vista operating systems. To do this, click Add Control → Voice Recognition, and then click the Config button. This opens a grid in which you can enter words or phrases to be recognized on the left, and add corresponding script commands on the right, dragged from a menu of possible commands at far right. Add commands like Forward, Back, Left, Right, and Stop, and you can now verbally control your robot (Figure L)! You can get creative by adding phrases such as "How are you?" and have the robot respond with *bleep bleep* noises, or use the operating system's

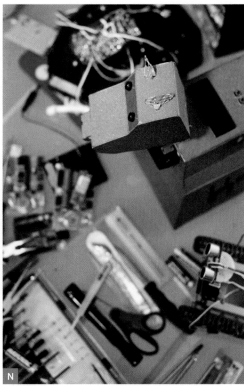

Colin Way

SMART DOGGIES

⬈ **Fig. M:** Formerly dumb Digger the Dog now moves autonomously, follows a red ball, and obeys voice commands, thanks to his EZ-B robot brain.

⬈ **Fig. N:** The author's award-winning, scratch-built K-9 robotic dog (faithful companion to TV's Doctor Who) was the basis for the EZ-B Robot Controller. K-9 avoids obstacles, follows people using onboard 3D mapping capabilities, synthesizes speech, and changes his personality based on human interaction.

built-in speech synthesizer to reply, "I am doing great, thank you for asking!"

If you have trouble with the voice recognition not hearing your voice, check the recording settings for your sound card in the control panel. The volumes for recording may be either too low or high. There is a wizard in Vista and Windows 7's control panel that you can use to set up the voice recognition. Those settings will be carried over to EZ-Builder.

Going Further with Your Bot
The reward of interacting with your first robot is exciting, and you can easily build on it by adding new sensors and EZ-Builder controls. Once you've tested each peripheral to confirm that it behaves correctly, you can create custom scripts to give your robot a unique personality. For example, you can script your robot to turn left if the distance sensor returns a value less than 8". The tutorials and community forum at ez-robot.com are great places to obtain assistance and share.

Because large corporations have no control over robotics, we're experiencing a great age of revolution. Today's hobby robots will, we hope, be refined into superior robots of the future. I created the EZ-Robot project to inspire and exercise your creativity, and I hope you'll share your creations with the community. The future of robots begins with your imagination!

🎥 See Digger and the other toy-robots in action at makeprojects.com/v/27.

DJ Sures is a roboticist who lives in Calgary, Alberta.

WELCOME, ROBOT OVERLORDS

Quadrotor Craze

Quadrotors, or quadrocopters, are all the rage in hobby robotics. MIT has built a fully autonomous quadrotor that uses an onboard Kinect to navigate and map its surroundings while generating a visual representation of where the robot has flown.

Algorithms scan image frames from the camera and recognize objects within the environment to assign depth and color to the models. The robot runs all the programs and controls in real time without the use of any additional sensors (makezine.com/go/quadrotor).

Sweden-based Daedalus Projects aimed to get a PCB with four motors to fly, the result of which is the amazingly tiny CrazyFlie, weighing in at 20 grams and boasting a 12cm×12cm footprint. Bonus points for technique: the accelerometer and gyroscope packages were soldered on using a frying pan (daedalus.nu/category/crazycopter).

Meanwhile, the Institute for Dynamic Systems and Control in Zurich, Switzerland, has developed a 35,000-cubic-foot Flying Machine Arena for testing mobile robots, most notably quadrotors exhibiting mesmerizing agility with synchronized choreography and cooperative ball juggling (www.idsc.ethz.ch/Research_DAndrea/FMA).

—*Nick Raymond*

Lab Lush

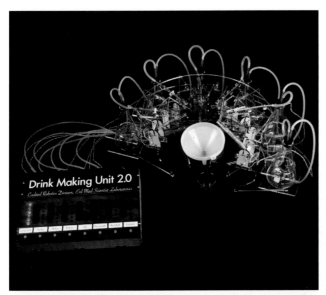

Evil Mad Scientist Labs strikes again, this time with its Drink Making Unit 2.0, a gorgeous hunk of machinery with parts sourced from such unlikely places as pet stores, chemistry labs, and Japanese gardens.

The DMU2 can dispense specific quantities of up to six liquids (Long Island Iced Tea, anyone?) and features a rad retro control panel. Precision bartending made easy (insert maniacal laughter)! makezine.com/go/drinks

—*Goli Mohammadi*

ON A ROLL

Orbotix has designed and built a robotic ball controlled by a driving app, or by tilting an iOS or Android device in the direction you want the ball to roll.

Sphero is essentially a smarter remote control toy: the ball collects data that can be used to add conditions to the gameplay, utilizing an accelerometer and gyroscope inside. In the app Tag, for example, players can pit their Spheros against each other. When the balls collide, they read contact data and compile game actions. And Sphero's open API lets developers build their own games and applications for it. makezine.com/go/sphero   —*Laura Cochrane*

OFF-ROAD SNAKE

Sriranjan Rasakatla's Semi Autonomous Snake Robot-p3 is a modular robot designed to move over rough terrain where wheeled robots fail.

In manual mode, the operator wears a data glove embedded with two accelerometers to intuitively send the snake commands. In autonomous mode, it utilizes an aerial-mounted camera to survey terrain and adjust its movements accordingly. Future applications include search and rescue missions and mine detection. makezine.com/go/snake

—*NR*

EMOTIONAL ROBOTIC BAND

Since the word "robotic" has a decidedly cold connotation, the idea of a temperamental robot rings my chimes. Created by the Scottish collective FOUND, Cybraphon may look like a 19th-century nickelodeon, but its guts are decidedly 21st-century.

A hidden computer runs custom software, while solenoids, servos, and infrared motion detectors play exotic instruments and antique machinery. Best of all, "performance is affected by online community opinion," leading it to act "just like a real band." Let's hope this review brings sweet music to Cybraphon's ears.

cybraphon.com — *Arwen O'Reilly Griffith*

Cavalcade of CoasterBots

Last year on makezine.com we ran a simple "CoasterBot" contest: build a bot using dead CD/DVDs (aka "coasters") as the main structural material and make it able to navigate a space. We taught CoasterBot basics, and our pals at Jameco offered a parts kit we put together.

We were floored by the variety and sophistication of the entries. Dan Ray's winning design, Jartron (seen here), walks, dances, sports stereoscopic rangefinder eyes, and has a laser nose that draws. You can see all the finalist entries, read the newsletter tutorials and blog progress reports, and still order the parts kit. makezine.com/robotbuild

— *Gareth Branwyn*

CELLPHONE FOR MACHINES

A wi-fi or Bluetooth module will connect to a robot locally, but the DroneCell can control it over any distance within the GSM quad-band cellular network coverage area — i.e. all over the world — and at altitudes of at least 10,000 feet above the nearest cell tower.

Connect the DroneCell's UART (universal asynchronous receiver/transmitter) TX and RX pins to a microcontroller or serial interface, and your project can send and receive text messages and jack into a standard cellular data plan. Hobbyists are using the versatile little communications portals to fly pilotless aircraft and relay data from weather balloons; industrial engineers are using them to outfit large-scale machinery for process control.

dronecell.narobo.com   —Paul Spinrad

PLANET H99

Learning how to program a robot can be challenging if you don't have an actual robot to run your programs on. Luckily, Carnegie Mellon University's Robotics Academy is developing a Windows-based robot-simulator game called Planet H99.

Users can program their robots in RobotC to control either a Lego NXT robot or a VEX robot. There's currently a working demo of the game available for download, and you can snag yourself a free copy of the final release by providing the Robotics Academy feedback on the demo.

makezine.com/go/h99   —Eric Chu

TurtleBot Hobby Platform

Looking for a reasonably priced personal robotics kit? TurtleBot from Willow Garage might be the answer, with kits ranging from $500 to $1,200.

TurtleBot uses open source software and builds on existing components like iRobot Create, Microsoft Kinect, and the ROS community's thousands of computer vision libraries.

There's no need to wire or solder anything — the TurtleBot comes with the only tool (a screwdriver) you'll need to assemble it. This leaves you the time and energy to concentrate on making a robot that can zoom around the house, build 3D images, take panoramas, and even deliver you food. I'll take fries with that!

willowgarage.com/turtlebot
—AOG

Hamster Inside

After seeing the wind-powered Gakken Strandbeest hacked to use solar and rubber band power and microprocessor control, I-Wei Huang, better known as CrabFu, was inspired with a delightfully silly idea: hamster power. "It's in so many what's-under-the-hood jokes," he says.

He removed the gearing and windmill and mounted a hamster ball instead; Meccano sprockets and chain were used to transfer the power from the hamster ball to the main crank. He then borrowed an acquaintance's pet hamster, Princess, to perform the inaugural run, much to the delight of his onlooking nieces, who now want their own hamster. "So," he says, "I may get some more test pilots in the near future." makezine.com/go/hamster

—LC

ULTRASONIC SENSORS

Have a project that needs distance sensing? Check out MaxBotix's line of ultrasonic sensors. Used by professional engineers, educators, and hobbyists, MaxBotix offers a wide variety of reliable and low-power sensors. With choices of wide to narrow beams, indoor and outdoor packages, and noise rejection on their higher end models, MaxBotix's sensors cover many demanding projects. maxbotix.com

—EC

GET A GRIP

JamBot, created by a University of Chicago team led by physicist Heinrich Jaeger, rocked the robotics world by approaching the challenges of gripping and picking up objects from a whole new angle: by using granular material encased in an elastic membrane. The mechanism jams onto an object, a vacuum is applied to tighten the grip, and any shape of object can be picked up. Maker Carlos Asmat Jr. documented a DIY version using a party balloon, ground coffee, and his lungs. Now that's genius. makezine.com/go/grip

—GM

BOTS ON THE ROCKS

Roboexotica has been partying with cocktail robots since 1999. Somebody needs to "document the increasing occurrence of radical hedonism in man-machine communication," right?

This collective of Austria-based makers hosts an annual festival devoted to bots that make drinks, complete with a symposium to explore the implications, an award for top bot, and workshops for those interested in building their own. What's not to love? roboexotica.org

—GM

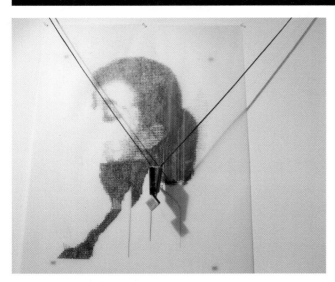

Art Hacks

Whether or not you think robots can make art, there's no denying the artistry of Harvey Moon's Drawing Robot. This wall-mounted, computer-controlled bot analyzes and re-creates images in pen. Ranging from intricate geometric patterns to beautiful portraits, the drawings resemble those of a futuristic Etch A Sketch and sometimes take weeks to complete.

Moon went to kickstarter.com to fund his second, more versatile prototype, and pledges exceeded his goal of $450 by almost $2,700. Check out more details and robot art at unanything.com. —*Craig Couden*

DIY UNDERWATER EXPLORATION

Dying for undersea adventures? Like robotic vehicles? Then check out the National Underwater Robotics Challenge, where student teams of all ages compete in mock undersea missions with DIY ROVs.

For more practical applications, check out OpenROV, a DIY telerobotics community. Group members built a spelunking rover equipped with an HD webcam and Arduino microcontroller connected to the surface via USB. openrov. com and h2orobots.org

—*CC*

ROBOTS OF MAKER FAIRE

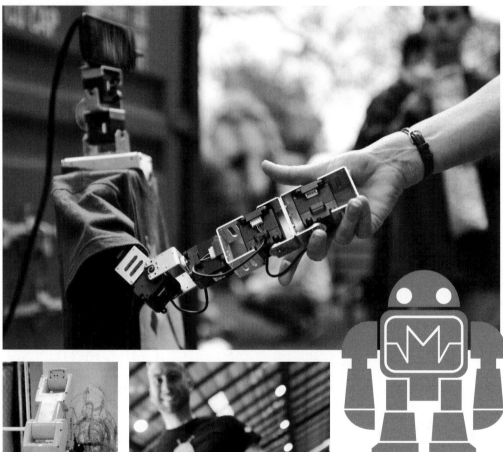

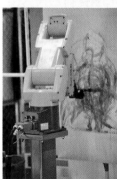

This year's Maker Faire Bay Area saw Google Android bots (like the smartphone-headed biped above), stolid industrial robots creating crazy art (à la Paint-Bot from *WALL-E*), and legions of DIY robots, from Mars rovers to bartenders to fierce combatants. See more great robots at Maker Faire in Detroit and New York this summer. makerfaire.com

—*Keith Hammond*

Gregory Hayes

Make: Projects

Build an aquarium whose gentle curtain of recirculating water sets the stage for an ever-changing jellyfish ballet. Then wire up a lightweight, portable sign that instantly displays anything you type in big, bright LED letters. Finally, turn back the clock and make the limestone-based spotlight that was state-of-the-art pre-Edison incandescent lighting.

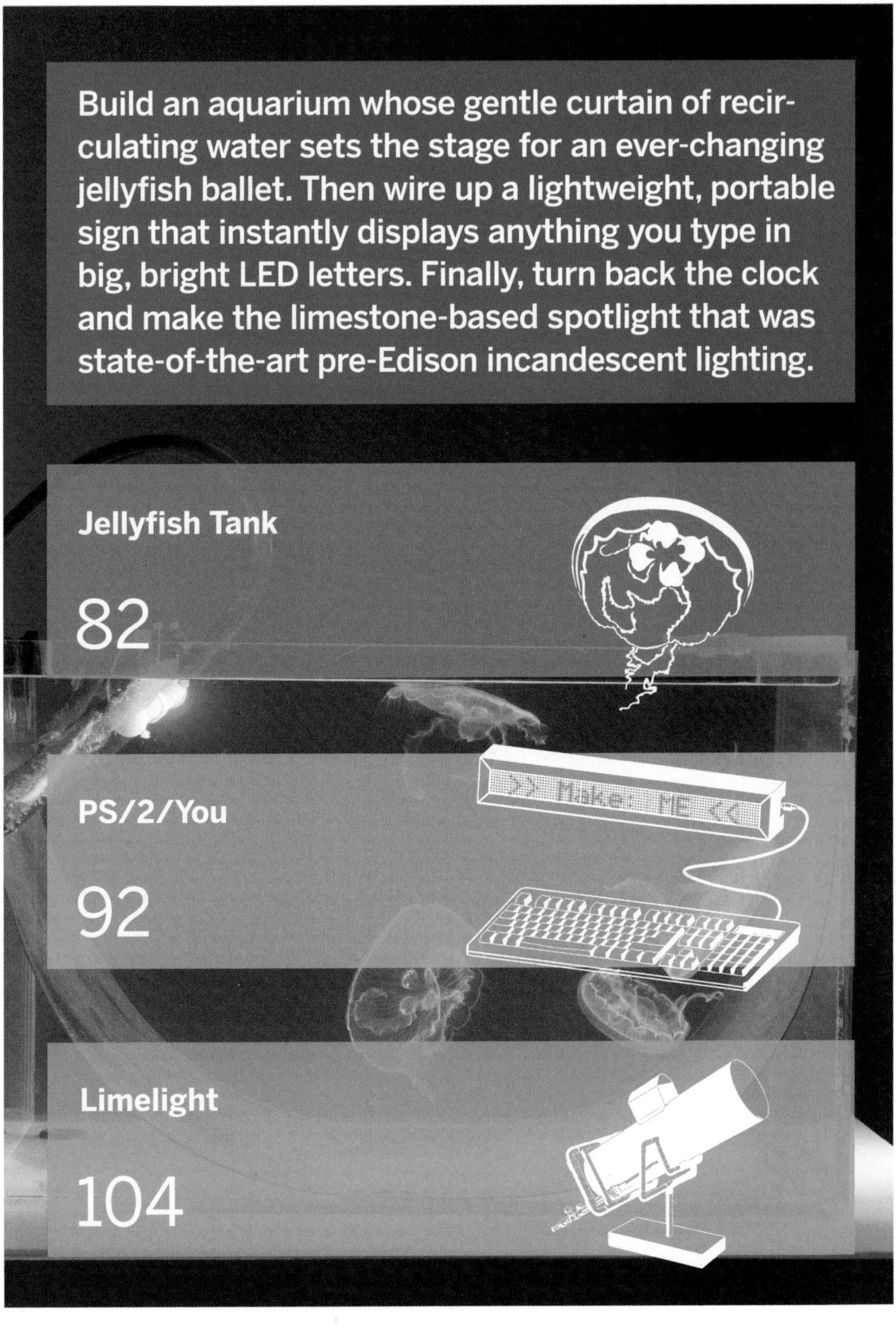

Jellyfish Tank

82

PS/2/You

92

Limelight

104

Garry McLeod

JELLYFISH TANK
Convert a regular aquarium into a jellyfish habitat.

By Alex Andon

I was always terrified of jellyfish as a kid. The thought of a slimy translucent blob just below the water's surface that could deliver a painful sting kept me out of the ocean most of the summer. It wasn't until years later when I collected a small jellyfish at the beach and observed it in an aquarium that I realized how stunningly beautiful they are. As it pulsed steadily and allowed its tentacles to flutter behind it, I was completely hypnotized. Soon, I decided to design and build an aquarium that could keep jellyfish alive and well.

Jellyfish can't live in a regular aquarium because they get sucked into the filtration pumps and liquefied. Using my extensive experience in building aquariums as part of my research projects at Duke and the University of Delaware, I developed a tank with a special water flow to keep jellies suspended in the middle of the tank.

The jellyfish aquarium design described in this article has no dead spots for water flow, eliminates strong points of suction, and creates a laminar water flow pattern that sweeps the delicate jellyfish away from the edges of the tank.

Alex Andon is the founder and president of Jellyfish Art (jellyfishart.com). He has a B.S. from Duke University in biology and environmental science.

| SET UP: p.85 | MAKE IT: p.86 | USE IT: p.90 |

Garry McLeod

JELLY JAR

Jellyfish require a special aquarium design to keep them from getting sucked into filtration intakes.

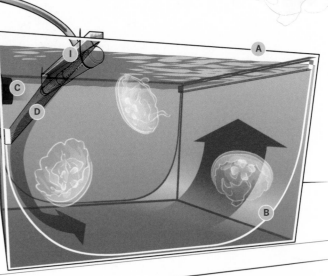

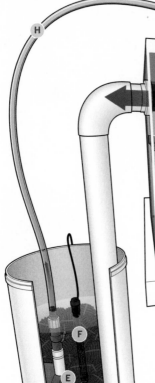

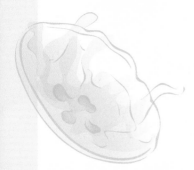

A Jellyfish swim around an aquarium tank, suspended safely in the relatively still water in the middle.

B The corners of the tank are rounded with a polycarbonate sheet to keep water flowing in a smooth, flat sheet along its sides and the bottom.

C A drain feeds water that exits the tank into an external bucket, for filtration and recirculation.

D A large exit screen keeps the jellyfish from getting sucked into the drain.

E Inside the bucket, a pump circulates the water.

F A heater in the bucket keeps the aquarium water at a consistent temperature.

G The bucket also contains high-surface-area plastic pieces. A helpful nitrifying bacteria called *Nitrosomonas* lives on these plastic pieces and digests the jellyfish waste. The pieces also trap particulates via filtration.

H Water is pumped from the bucket back into the tank via a flexible hose and the spray bar.

I Water re-enters the tank through a spray bar that washes the water in a flat sheet across the outside of the exit screen. This prevents the jellies from getting stuck to the screen.

James Provost

SET UP.

A

B

C

D

E

F

G

MATERIALS

A. Polycarbonate sheet, 1/32" thick
width 1" less than the interior width
of your tank, and length equal to the
tank height plus the tank length

B. Salt water pump able to pump at
least 5 times the volume of your tank
per hour at 0 head pressure

C. PVC glue

D. PVC pipe and fittings:
» PVC caps, slip, 1/2" (2)
» PVC pipe, Schedule 40,
1/2" diameter, 2' length
» PVC tee fitting, slip × slip
× female pipe thread, 1/2"
» PVC pipe, Schedule 40,
1½" diameter, 2' length
» PVC hose barb fittings,
male pipe thread (2) sized
for 1/2" flexible hose
» PVC ball valve, slip × slip, 1/2"
» PVC fitting(s) to connect 1/2" hose
barb to 1/2" valve to water pump
This will depend on your pump. My
pump had a male pipe thread (MPT)
fitting, so I used two adapters: female
pipe thread (FPT) at both ends, and
MPT × slip. It would have been simpler
to use just one slip × FPT adapter.

» PVC adapter, male pipe thread
× slip, 1½"
» PVC elbow, slip × slip, 1½"

» Plastic epoxy
» Teflon tape
» Threaded bulkhead fitting, 1½"
» Flexible hose, 1/2" diameter,
3' length
» Mesh screen, cut to (width of tank
+ 2") × (1/2 height of tank)
» Plastic bio-balls, about 3gals
or plastic pieces of any kind with a lot
of surface area, such as green plastic
army men
» Twist ties or zip ties (2)
» Acrylic strips: 1/2" × 1" × outside
width of tank; and 1/2" × 1" × inside
width of tank (2) for braces to hold
the screen and polycarbonate sheet
» Salt water
» Submersible aquarium heater
50W for up to 15gals; 100W for up
to 25gals; 200W for up to 50gals.
For cold-water species, replace the
heater with a drop-in water chiller.
» Aquarium tank At least 4 gallons and
approximately cube-shaped. I built a
23"×28"×17" acrylic tank (see Tools).

TOOLS

E. PVC pipe cutter

» Dremel with 1/8" drill bit and router
» Tape measure
» Diamond hole saw (optional)
if you're using glass tank

***To build your own 28"×23"×17"
(L×W×H) aquarium (includes polycarbonate and acrylic listed above):***

F. Acrylic cement

G. Aquarium-grade silicone caulking
available at aquarium or pet stores

» Acrylic sheets, 1/2" thick: 28"×23"
(bottom); 28"×17" (2, sides);
22"×17" (2, ends); 22"×1" (2,
braces); 23"×1" (brace) Buy cut-to-
size from a local plastics distributor
or cut yourself (carefully) using a
table saw with carbide blade in a well-
ventilated area.
» Polycarbonate sheet:
21⅞"×45"×1/32"
» Dimensional lumber, 2×4,
8' lengths (2) to build a 90° jig
» Ruler or straightedge

Gregory Hayes

MAKE IT.

BUILD YOUR JELLYFISH TANK

Time: 3–5 Days
Complexity: Easy

1. BUILD THE TANK (OPTIONAL)

1a. The acrylic sheets come with a protective wax paper covering. Peel back the paper 1" from each edge you plan to glue, and use a ruler or straightedge to tear the excess (peeled) paper off.

1b. Use a saw and drill to build a 90° jig from the 2×4s to hold the acrylic pieces in place while the glue sets.

1c. Use acrylic cement at the joints to weld the tank together (refer to instructions included with the cement, and see Step 3a for placement of braces). The cement enters the cracks through capillary action.

1d. Use aquarium-grade silicone caulking to seal all the glued joints to prevent water from leaking (normal silicone caulking isn't rated for salt water and will degrade quickly; it may also leach into the water and kill your jellyfish).

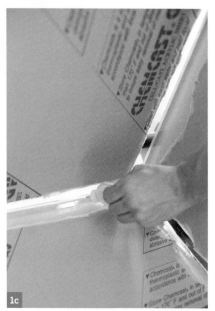

1e. After the joints are completely dry, peel the rest of the wax paper off the acrylic.

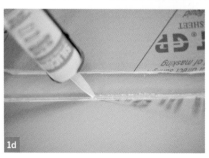

2. ASSEMBLE THE SPRAY BAR

2a. Cut the ½" PVC pipe into 2 equal pieces — these will be assembled with the tee fitting and 2 slip caps to form the spray bar.

2b. Glue the 2 pieces of ½" PVC pipe on either side of the tee fitting. Glue the slip caps onto the ends using PVC glue.

2c. Using a Dremel or a drill press with a ⅛" bit, drill holes in the PVC pipe on the side opposite the tee fitting's opening. The holes should be approximately ½" apart, following a straight line from one end of the spray bar to the other.

2d. Screw a hose barb fitting into the tee using teflon tape.

3. MAKE THE EXIT SCREEN

3a. Using plastic epoxy or acrylic cement, glue the long acrylic brace on top of the tank at a distance from the drain side of the tank equal to ⅓ of the tank height.

3b. Cut out the corners of the screen as shown. The longer, straight sides will be glued to cross-braces and the short, beveled sides will be glued to the inside of the tank.

2a

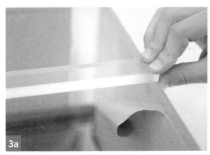

2b

2d

3a

3b

TIPS: The PVC pieces should be cut so that the final length of the spray bar is a little narrower than the width of your tank.

Before gluing, make sure the assembled spray bar fits inside the width of your tank.

2c

3c. Epoxy one long side of the screen to one of the shorter acrylic braces. Add more epoxy on top, and sandwich the screen between the brace and the inside of the tank, on the drain side, about ⅓ of the way down from the top. Hold it in place and allow the epoxy to cure to a strong hold.

3d. Glue the opposite side of the screen under and around the long brace on top of the tank. Gently pull the screen taut and hold it in place while the epoxy cures.

Fold the short sides of the screen in and epoxy them against the inside of the tank.

4. MAKE THE DRAIN

4a. Cut out a circle slightly bigger than the bulkhead fitting in the middle of the tank's exit screen side, as far up as it can go without the fitting overlapping the top of the tank.

NOTE: If you're using an acrylic tank you can make the hole with a hole saw or Dremel with a router fitting. If you're using a glass tank you need to make the hole with a diamond hole saw.

4b. Install the bulkhead fitting in the hole.

Using teflon tape, screw the 1½" male pipe thread (MPT) × slip adapter into the outside of the bulkhead fitting.

4c. Cut a 3" long piece of 1½" PVC pipe and use it to connect the 1½" male pipe thread MPT × slip adapter to the 1½" PVC elbow, pointing down.

4d. Cut a piece of 1½" PVC pipe long enough to run from the 1½" PVC elbow into the bucket beside the tank.

5. ASSEMBLE THE BUCKET

5a. Using the necessary adapters, attach the ½" hose barb fitting to the ½" valve, and attach the valve to the pump.

5b. Attach the ½" flexible hose to the hose barb fitting. Put the pump and heater at the bottom of the bucket, and fill the bucket with bio-balls or whatever pieces of plastic you'll be using for the filter.

5a

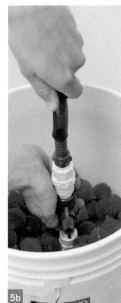
5b

6. ASSEMBLE THE SPRAY BAR

6a. Attach the free end of the ½" flexible hose to the hose barb fitting on the spray bar.

6b. Use 2 zip or twisty ties to tie the spray bar in place so the sheet of water will wash across the exit screen.

6a

6b

7. INSERT THE FLOW SHEET

7a. Epoxy the remaining acrylic brace flat inside the end of the tank opposite the drain, flush with the top.

7b. Wedge the polycarbonate sheet into the tank so that it runs along the bottom and curves up at each end, held down by the two braces glued inside the tank. (If you built a tank 28"×23"×17" high, your polycarbonate sheet should measure 45"×21⅞".)

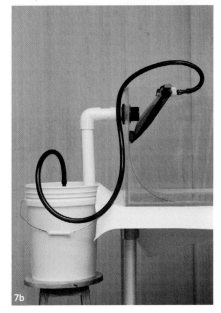
7b

7b

USE IT.

JUST ADD JELLYFISH

Fill the tank with salt water. You can get pre-mixed salt water or mix it yourself with aquarium salt, but make sure you use distilled or reverse-osmosis (RO) filtered water. Do not use unfiltered tap water.

Plug in the pump and heater, and allow the water to heat to the natural temperature of the jellyfish you'll be adding (almost all commercially available jellies are kept at 77°F).

Add jellyfish. Moon jellyfish do best in these aquariums. (If you can't collect jellyfish from the ocean, you can buy jellyfish and jellyfish food from my company: jellyfishart.com). Adjust the valve so there's just enough water flow to keep the jellyfish from settling on the bottom. Water has a lot of momentum, so

> ◤ TIPS: Change 25% of the water every 2 weeks once your tank is running. For maximum longevity with coastal jellyfish (from their natural lifespan of about 8 months up to a year or more), feed them twice a day and use a robust filtration system.
>
> Try lighting your jellies from the side or top of the tank. The light will bounce off their translucent bodies and make them glow. You can use color-changing LEDs to make the jellyfish light up with whatever colors you shine on them.

wait at least 10 minutes after adjusting the valve to observe the effect on water flow. Adjust the position of the spray bar so water washes over the exit screen without creating any air bubbles, and the jellyfish don't get stuck to the screen.

You can feed your jellyfish live brine shrimp or frozen plankton foods made specifically for jellyfish. Learn more about the care and feeding of jellyfish at jellyfishart.com. ◪

Garry McLeod

PHOTOGRAPHING YOUR JELLYFISH

By Steve Haddock

Taking great photos of jellyfish comes down to controlling 3 things: the tank, lights, and camera. These tips will help you reveal the stunning beauty of your jellies through photography.

As with all creative processes, these rules are made to be broken as your artistic impulses dictate. Motion-blurred images of softly focused jellies can be as beautiful as crisp shots on a clean black background.

If you do collect wild jellies and get good pictures, be sure to submit them to jellywatch.org!

Tank

For a black background, put a piece of velvet behind the tank, or put a piece of black acrylic inside the tank in the back.

For illuminated backgrounds, you can play with different colors, but you'll want to use a low depth of field to make sure the background isn't in focus. You can use a second light source to illuminate the background, so that you can control the exposure independently.

Camera

You can use either a fancy SLR or a cheaper compact camera. If you want a dark background, set the exposure manually — most cameras will try to expose to a uniform gray. If your camera doesn't have manual exposure settings, then you'll want to set the exposure compensation to −3. This will also make a shorter exposure to avoid motion blur.

Set the focus manually on the SLR, and depending on the shot, use macro mode on the compact. With an SLR, initially set the focus so the jelly is the right size in the frame. Then, instead of constantly turning the focus ring, move the camera closer or farther from the tank to get the subject into focus.

You'll typically get better results if you underexpose the images relative to what you think they should be based on the camera's preview. You can't recover overexposed areas, but you can bump up the levels of underexposed shots.

Depth of field in macro shots can be a problem. If you can set the aperture, the higher numbers will give you greater depth of field (more things in focus).

Lights

An external light source is the most important element — you can't use an on-camera flash to get good pictures.

You can use a strobe or a clip lamp as a light source. Position it so it shines in from the side of the tank, farther from you than the plane of the front window. If you're using a clip lamp, set your camera's white balance manually to "incandescent" (the light-bulb icon). Otherwise your pictures will all have a yellowish tint.

Make a cover out of cardboard (for the strobe) or foil (for the lamp), to narrow the beam of light. Ideally, the slice of light will illuminate just the jellyfish, and not the back of the tank or the front surface (or you, reflected in the front). Turn off the room lights while taking pictures, and use a flashlight or the clip lamp to focus. This will prevent capturing your reflection in the glass. ◪

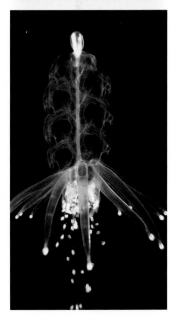

Steve Haddock is a marine biologist with an engineering background who studies jellyfish and bioluminescence at the Monterey Bay Aquarium Research Institute. He co-wrote *Practical Computing for Biologists* (practicalcomputing.org).

Steve Haddock

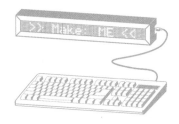

PS/2/YOU
Go-anywhere, instantly updatable glowing digital message board.

By Immanuel McKenty

It all started with a small LCD salvaged from an old printer. I recruited my code-savvy older brother, Adam, and we soon had the LCD displaying text from an Arduino microcontroller. This was neat, but it was inconvenient having to plug the Arduino into a computer for reprogramming whenever we wanted to change the text.

We needed something for inputting the text, and it didn't take long to find a PS/2 keyboard code library for Arduino — which confirmed my observation that anything that communicates with wires has probably been hooked up to an Arduino. I salvaged a PS/2 port from an old computer motherboard, and after some trial and error, we could plug in a common PS/2 keyboard ($5 new) and type messages directly into the Arduino and out to the LCD.

The LCD was so small, however, that hardly anyone noticed our witty remarks. We needed a bigger display. After looking at many appallingly priced commercial LED matrix products, we found a new and much cheaper offering: Sure Electronics' 8×32 display boards. They cost $9 each and you can cascade up to four into one long display. We ordered three, and by the time they'd arrived, the Arduino community had already produced a library to run them. (Our code is based on two open source Arduino libraries: *PS2Keyboard*, by Christian Weichel; and *MatrixDisplay*, by Miles Burton.)

The result is our PS/2/You system, which displays keyboard-typed messages in 2"-tall LED letters that can be read from quite a distance. You can store and switch between six different lines of text, and it automatically scrolls through lines that are too long for the display. Power comes from an AC adapter or six AA batteries for portable operation, and the whole thing is housed in a sturdy wooden frame.

Immanuel McKenty is an 18-year-old home-schooler. Among other things, he rather enjoys taking things apart, and sometimes even manages to get them back together again.

SET UP: p.95	MAKE IT: p.96	USE IT: p.103

GET THE KIT

makershed.com
/ps2you

COLLABORATE ON

Make: Projects
makeprojects.com

LOOK, MA — NO COMPUTER!

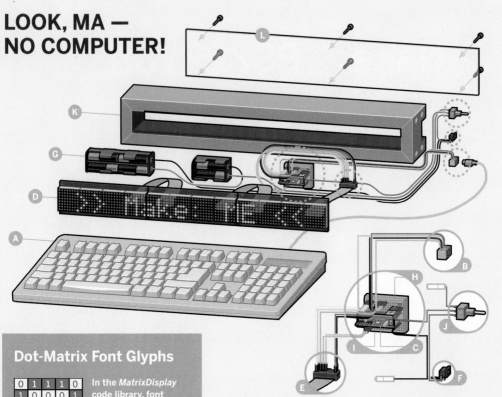

Dot-Matrix Font Glyphs

0	1	1	1	0
1	0	0	0	1
1	0	0	0	1
1	0	0	0	1
1	1	1	1	1
1	0	0	0	1
1	0	0	0	1
1	0	0	0	1
1	2	3	4	5
7F	88	88	88	7F

In the *MatrixDisplay* code library, font glyphs for the PS/2/You are stored in a multidimensional array of hexadecimal values (base 16). Each character of the font corresponds to one line in the array.

As an example, one line looks like this:

(0x7F, 0x88, 0x88, 0x88, 0x7F),

This sequence represents the capital "A" character. Convert the hex values into binary, write them vertically from top to bottom, and you get a 5×8 array of 1s and 0s. Convert these bits to On and Off LEDs, and you'll see your big "A" up in lights.

The PS/2/You code actually only generates 5×7 characters, leaving the top row of LEDs unused — 5×7 is a standard size for LED matrix characters, and it allows more characters to fit on the display without looking overly elongated.

Hexadecimal	7F	88
Binary equivalent (1×8 LED column)	01111111	10001000
Decimal equivalent (not that it matters)	127	136

A With each key press and release, a **PS/2 keyboard** sends a "scan code" identifying the key over its data pin (pin 1).

B The keyboard plugs into a **PS/2 port**, which breaks its pins out into connectable contacts.

C The PS/2 port connects to an **Ardweeny microcontroller**, an Arduino clone, which interprets the scan codes. For character key inputs, the software looks up the corresponding 5×7 glyphs, then writes them via ribbon cable. The Enter key toggles between Input and Output modes.

D Three 8×32 **LED modules** display the glyphs sent from the Ardweeny. Data is transmitted serially over a single wire.

E **Ribbon cables** carry outputs from the Ardweeny to the display modules one by one, in a "cascade" configuration that acts as one 8×96 pixel display.

F A 9V–12V AC wall adapter provides plugged-in power.

G Six **AA batteries** power the display for on-the-go operation.

H A **voltage regulator** and **filtering capacitors** pare the voltage from batteries or wall wart to the circuitry's required 5V.

I A **mini solderless breadboard** holds and connects all the circuit components and wires.

J An **SPDT (single-pole, double-throw) switch** selects battery power, plug-in power, or power off.

K A custom **wood enclosure** holds all the components.

L A **plexiglass back cover** keeps the components inside while letting curious spectators marvel at the geeky goodness.

Rob Nance

SET UP.

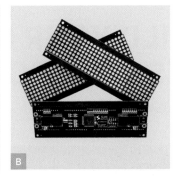

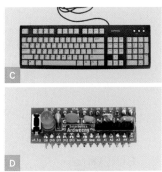

MATERIALS

See makeprojects.com/v/27 for suppliers, prices, and other sourcing information.

A. MAKE PS/2/You Kit item #MSPS2 from the Maker Shed (makershed.com/ps2you), includes items B–I below and all other electronic components, except batteries.

B. Dot matrix LED display modules, 8×32, with ribbon cable (3) Sure Electronics item #DE-DP106, about $9, or equivalent module. This item was recently discontinued but is available on eBay; see makeprojects.com/v/27 for alternative modules, drivers, and code.

C. Computer keyboard with PS/2 connector They're readily available at thrift stores. You can also use a USB keyboard with a USB-to-PS/2 adapter.

D. Ardweeny microcontroller This small, cheap Arduino clone fits into a standard 14-pin DIP socket, but it doesn't come with an onboard 5V voltage regulator or an FTDI USB-serial converter for programming.

E. 5V voltage regulator You can use a 7805, but the low-dropout LM2937

will make your batteries last longer, especially with lower-voltage NiMH AAs.

F. PS/2 port from an old PC motherboard; ask your local computer shop.

G. Solderless breadboard, self-adhesive, mini

H. Breadboard jumper wires (around 20, multiple colors), or solid core 22AWG wire Jumpers are easier to use and well worth the expense.

I. FTDI serial programmer such as the FTDI Friend, Maker Shed #MKAD22, $15

» **9V–12V AC wall adapter** can be found for $1–$2 at most thrift stores
» **DC power jack to match your adapter** probably a standard 5.5mm/2.1mm barrel jack
» **Power switch, SPDT (on-off-on)**
» **Capacitor, ceramic, 0.1µF** labeled "104"
» **Capacitor, electrolytic, 10µF**
» **Batteries, AA (6)**
» **Battery holder, 2×AA**
» **Battery holder, 4×AA** in a long, flat, 2×2 configuration
» **Stranded wire, 22AWG, 4' total** We used red and black.
» **Electrical tape or heat-shrink tubing**

» **Acrylic/plexiglass sheet, clear,** ⅛" thick, 21"×4" Lexan will work great but is more expensive.
» **Wood screws, #8 flathead,** 1¼" long (8)
» **Wood screws, #8 pan head,** ½" long (6)
» **Wood screws, #6 pan head, ½" long (12, optional)** see page 97
» **Dimensional lumber, 1×4** (¾"×3½"), 4' length, or 1×2 (¾"×1¾"), 8' length see page 97

TOOLS

» **Measuring tape or long ruler**
» **Handsaw or chop saw**
» **Table saw (optional)** see Step 1g
» **Chisel**
» **File**
» **Hammer or mallet**
» **Drill and drill bits:** ⁵⁄₆₄", countersink
» **Screwdriver, medium**
» **Soldering iron and solder**
» **Desoldering braid or solder sucker**
» **Wire cutters**
» **Needlenose pliers (optional)** handy for plugging in breadboard jumpers
» **Multimeter**
» **Glue gun and hot glue**
» **Computer with internet connection and USB port**

Gregory Hayes

MAKE IT.

BUILD YOUR DIGITAL MESSAGE BOARD

Time: 3–4 Hours
Complexity: Moderate

1. BUILD THE FRAME

1a. Cut the 1×4 lumber in half lengthwise to make 2 strips about ¾"×1¾" (a nominal 1×4 is actually around ¾"×3½"). Use a narrow-kerf blade if possible.

NOTE: This frame fits the DE-DP106 LED modules; if you use alternative modules you'll have to improvise a frame to fit them.

1b. Line up the 2 boards beside each other on a flat surface with their narrow edges up. Place one of the display panels facedown between the boards, so that the flanges on the panel rest on the boards, with the pro-truding LED matrix between them. Gently squeeze the boards snug against the sides of the LED matrix, and measure between the outside edges of the boards. This measurement is the length of the frame's end pieces.

1c. Use a chop saw or hand-saw to cut a 45° angle on one end of each piece, oriented so the cut goes diagonally across the narrow edge. Measure 18¼" down the board's length from the inner edge of the cut, and make a second 45° cut that angles back out. Repeat on the second board. These will be the 2 long sides of the frame.

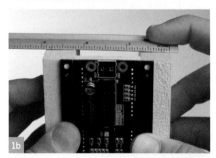

NOTE: The frame style isn't crucial, so let your creativity (and materials) have a say in the design. I had a woodshop at my disposal, so I made some-thing like an extra-deep picture frame with mitered corners and a slot cut in the long sides to hold the display panels.

Immanuel and Adam McKenty

1d. On each of the leftover board pieces, mark the distance measured in Step 1b down from the sharp, outside edge of the miter cut, along the longer face of the board. Cut one at 45° angled *in* from the measured length (the mirror image of the first cut), which will be each piece's longest dimension. Don't cut the other piece yet.

1e. Desolder the PS/2 port from its donor motherboard. Line up the DC jack, PS/2 port, and power switch atop the edge of the marked but uncut short piece of wood. Mark out a notch in the edge of the board just wide enough for all of them to fit next to each other and deep enough that the tallest component will sit flush.

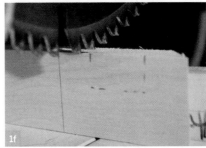

1f. With a handsaw (or chop saw) cut the 2 edges of the notch. Make a few cuts to the correct depth in the middle of the notch, then chisel out the rest of the wood and file the bottom of the notch smooth. The ports and switch should slide into the notch easily but without extra space. Cut the second (notched) end piece as marked in Step 1d.

1g. Set the blade of your table saw to a depth of ⁵⁄₁₆" (the size of the flanges on the display panels). Cut a groove lengthwise down the inside face of each long frame piece, ¼" in from the edge (preferably with a narrow-kerf blade).

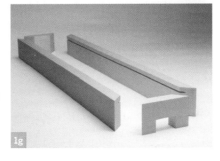

NOTES: The panels should be able to slide into the slots and be roughly flush with the front of the frame.

If you don't have access to a table saw, another way to make the frame is to cut two 1×2 boards (really ¾"×1½") to 18¼", and two shorter lengths of 1×2 to fit on the ends. Then, rather than cutting grooves for the display module flanges, use 12 #6 screws to secure the 3 modules onto the front of the frame through their pre-drilled mounting holes.

1h. Slide the 3 display panels into the slots in the long frame pieces, all oriented so that the writing on the printed circuit boards reads right side up as you look into the back of the frame. Fit and hold a frame end piece in place at each end, then drill ⁵⁄₆₄" pilot holes and countersinks for the 1¼" #8 screws that hold the frame together.

The notched end belongs on the left side of the frame as you read the circuit board backs, which is the right side as you view the front of the display. Put the screws in as you go to keep the frame together. Ensure that your countersinks are deep enough, and (to avoid splitting the frame pieces) don't overtighten the screws. The frame is now finished!

2. CUT THE BACK COVER

2a. Place the assembled frame on top of the plexiglass sheet. Use a screw or other sharp object to scratch a mark around the edge of the frame.

2b. Cut out the back cover with a handsaw, table saw, or the cutting implement of your choice. Line up the cut piece on the back of the frame and drill 6 pilot holes through the back cover material and into the frame itself. It's now ready to be closed up once all the electronics are in place and functioning.

TIPS: Be sure to install the screws on the bottom edge of the frame at least ⁵⁄₈" up from the bottom to avoid contact with the screws we'll use to secure the back cover.

If you'd like, add some glue to the notched end of the frame for extra strength, but leave the other end unglued so you can unscrew it and slide the display panels back out.

3. WIRE THE POWER AND PS/2 PORTS

3a. Wire the 2 battery packs in series by soldering the red (+) lead of one to the black (−) lead of the other.

3b. Position the battery packs in the frame (I put them at the end opposite the notch). and lengthen the remaining 2 wires if necessary by splicing in stranded wire to let them reach the notch, where you'll connect them to the DC jack and power switch. Insulate connections with electrical tape or heat-shrink tubing.

3c. Cut one end off a black breadboard jumper, and strip and tin the wire. Repeat with a red jumper. Solder the cut end of the black jumper and the black wire from the battery packs onto the DC jack's negative terminal. Solder a short chunk of stranded red wire between the DC jack's positive terminal and one of the outside contacts on the switch. Solder the cut end of the red jumper to the switch's common terminal, and the battery positive to the free outside switch terminal.

3d. Cut one end off 4 more breadboard jumper wires: red, black, blue, and white (or your equivalent). Strip, tin, and solder them onto the positive, negative, data, and clock pins on the PS/2 port (see pin diagram), and use a multimeter set to "continuity" to confirm the pin-wire connections.

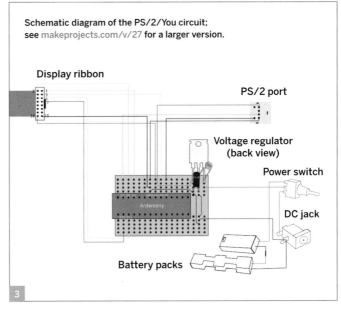

Schematic diagram of the PS/2/You circuit; see makeprojects.com/v/27 for a larger version.

Display ribbon

PS/2 port

Voltage regulator (back view)

Power switch

DC jack

Ardweeny

Battery packs

3

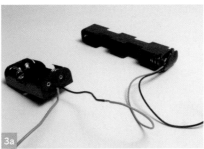

3a

NOTES: The battery packs need to be removable for battery changes, so you can secure them in the frame using velcro, although the back cover and display panels seem to hold them in place nicely.

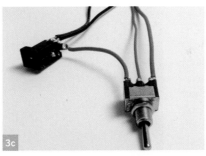

3c

If all went well, your switch should have an off position in the middle, a battery power position to one side, and adapter power on the other.

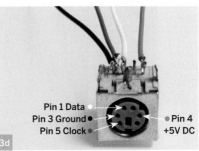

Pin 1 Data
Pin 3 Ground
Pin 5 Clock
Pin 4
+5V DC

3d

I used blue for data wires, white for read/write and clock wires, yellow for the display panels' "CS" wires, and red and black for power and ground — although I accidentally switched blue and white here on the PS/2 port. Having a consistent color scheme will make it much easier to figure out what's going on.

3e. Dry-fit the PS/2 port, DC jack, and power switch again into the notch at the end of the frame. Heat up a glue gun and dab a bit of glue on each one before quickly pressing it tightly into the notch.

NOTE: This is admittedly an unorthodox way of attaching what would normally be PCB or panel-mounted components, but it's very strong and relatively tidy — a good substitute when there's no PCB or mountable panel nearby.

4. CONFIGURE THE DISPLAY PANELS

4a. Each LED display panel comes with a ribbon cable and has 2 ports on its backside that the cable will connect to.

Use 2 of the ribbon cables to chain the 3 panels together end-to-end by their adjacent ports. Plug one end of the remaining cable into the port closest to the switch and power jack. Fold this ribbon up tightly and use hot glue to attach its free end plug to an inner side of the frame, with its holes facing up toward the back of the frame.

TIPS: Make sure the plug holes have enough headroom to let you plug in breadboard wires without their scratching the transparent back cover (once it's screwed on). Also make sure there's enough room between the glued-in plug and the board's other ribbon cable port to accommodate the mini solderless breadboard.

4b. Each LED display panel has a block of little DIP switches labeled CS1, CS2, CS3, and CS4. These switches determine how the microcontroller identifies each panel. The PS/2/You code numbers the displays left to right, looking from the front, so turn off all but switch 3 on the panel nearest the notch, all but switch 2 on the middle panel, and all but switch 1 on the panel at the battery end.

(To see what these switches do, set them to some other sequence once you have your display up and running.)

5. ADD THE ARDWEENY

5a. Plug your Ardweeny into the breadboard straddling the trench, with the green LED near the top, taking up the first 14 rows. Plug the voltage regulator into the bottom 3 rows on one side.

Plug the 0.1µF capacitor in between the voltage regulator's input and ground legs (typically the sequence is IN-GND-OUT, going left to right looking at the front of the regulator, but check your datasheet to be sure).

Plug the 10µF capacitor's positive leg in the regulator's output and its negative leg (marked with a stripe) into the regulator's ground bus.

5b. Peel off the breadboard backing (exciting!) and stick it onto the flat, surface-mount Holtek chip on the back of the display panel closest to the switch and ports.

5c. Since the voltage regulator's ground bus is getting crowded by now, use a small jumper to extend it to an unused bus on the other side of the breadboard.

Plug the red power wire from the switch's common into the regulator's input bus, and the GND wire from the DC jack into the new ground bus. Use a red jumper to connect the regulator's output to the Ardweeny's power (labeled "+") and a black jumper to hook the ground bus to Ardweeny GND.

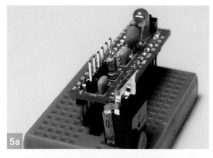

The ribbon cables each have 15 wires, and their plugs have 2 rows of holes that connect to odd-numbered wires along the top and even-numbered ones along the bottom.

5c, 5d

5e

5d. Plug the PS/2 port's power wire into the Ardweeny's power and its GND into the Ardweeny's GND bus. Plug the PS/2 port's read/write wire into Ardweeny pin D3 and its data wire into Arwdeeny pin D7.

5e. Use jumpers to connect your components to the ribbon cable plug glued inside the frame. Note that the odd-numbered pins on the 2×8 plug run along the side with the small bump, opposite the side that the ribbon comes in.

To begin, connect CS2, the first wire on the ribbon (marked in pink) to Ardweeny pin D5. Connect CS3, the second wire, to D6, and ribbon wire 3 (CS1) to Ardweeny D4. For the display's read/write input, connect ribbon wire 5 to Ardweeny D11, and for the data input, connect wire 7 to Ardweeny D10.

Finally, connect ribbon wire 15 to Ardweeny ground, and ribbon wire 16 to Ardweeny power. (You can also wire power and ground to the buses off the voltage regulator.)

6. PROGRAM IT

6a. If you haven't already, download and install the Arduino IDE (integrated development environment) from arduino.cc/en/Main/Software. Launch the IDE.

To configure it for the Ardweeny, which acts just like an Arduino Duemilanove, click the menu item Tools → Board → Arduino Deumilanove or Nano w/ATmega328.

Then tell it which USB port you'll program the Ardweeny through by clicking Tools → Serial Port and selecting the highest-numbered COM port (if there's more than one).

6b. Download the code package *PS2You_code.zip* from makeprojects.com/v/27 and unzip it.

Move the *PS2Keyboard* and *MatrixDisplay* folders into your Arduino libraries directory.

6c. Restart the Arduino IDE and open up the code file *PS2You.pde*. Connect your computer to the Ardweeny with the FTDI programming adapter. Click Verify and Upload, and if all is well, a moment later the display will light up with the default text.

6d. Unplug the programmer, load some batteries into the battery packs (if you're going to be using battery power), and screw on the back cover. You're all set!

MAKE SOME BRIGHT REMARKS

Input and Output Modes

The PS/2/You has an Input mode for entering text and an Output mode for displaying it. Pressing any alphanumeric key puts the system into Input mode, and hitting Enter gets you back to Output mode.

In Input mode, the PS/2/You displays a single line of text as you type it in, a maximum of 100 characters long. You can store up to 6 lines of text (this number is settable by changing the value of numLines near the beginning of *PS2You.pde*). Use the up and down arrow keys to select which line to edit, and backspace over any line of text or hit Escape to delete it.

In Output mode, the display loops through the stored text lines on its own, displaying each for one second, or if the line is longer than 16 characters, it will scroll across the display before moving onto the next line. If only one line of text is stored, it displays continuously.

Text Messaging

The uses for this contraption are many. Plug the keyboard in and enjoy putting your wittiest "wiseclacks" on it in the safety of your home, shop, or office — or use the battery option to take it into the wide world. We like to have the keyboard accessible so that passersby can add a riposte or two to the dialogue. But if monologue is more your thing, you can always keep the keyboard out of reach.

Add a dowel as a removable handle so you can wander the streets digitally promoting your geekified political leanings. Keep score (or heckle) at sporting events, deliver birthday greetings, advertise your wares at a farmer's market, beam cryptic messages to your neighbors — the possibilities are endless!

That's Illogical

De Verdad!

Hello Geek Girl

Wanna Draq?

Fear the Beard!

Further Illuminations

There's plenty of room for improvement to the code. Try using Ctrl and other keys to modify how the text displays: flashing, sliding in from the top, fading in, or other effects.

Four display panels can be cascaded together for a longer display, and Sure Electronics sells an identically programmable 8×32 panel with 5mm instead of 3mm LEDs, so a jumbo PS/2/You is almost inevitable.

Roll Your Own Glyphs

The display font is defined by hexadecimal values in the *font.h* file. It's not user-friendly for editing, but Brent Morse has made a free applet that lets you design your own 5×7 LED display glyphs (morse-code.com/id89.htm). Use it to modify the font, or make custom smilies or any other pattern you like. ◪

➕ Visit makeprojects.com/v/27 for the PS/2/You circuit schematic, code, and more.

LIMELIGHT
Experience pre-Edison incandescent lighting.

By Peter Tabur

When I was a wee lad, I enjoyed the television show *Connections*, by James Burke. Each episode traced the technical developments that ultimately resulted in some modern marvel. For example, the 18th-century theory that disease was caused by "bad air" ("mal'aria" in Italian) inspired Alessandro Volta to create a fanciful pistol-shaped sparking device that checked for methane. This led to spark plugs and ultimately to the automobile.

Burke also discussed the limelight, which came to prominence in the mid-1820s when it was used by land surveyors in Ireland who needed a brighter light to sight between mountain peaks in murky weather. Atop the mist-shrouded Slieve Snaght, Thomas Drummond heated a ball of lime in a flame of burning alcohol fed by an oxygen jet. The resulting incandescence glowed far brighter and whiter than any bare flame, making it visible from Divis Mountain, more than 66 miles away. Limelights fueled by oxygen-hydrogen combustion subsequently enabled bright spotlights in theaters, which is mainly how the word "limelight" survives today.

The idea of actually building a limelight became a splinter that lodged in my brain for 30 years, and last year I finally did it. My first experiment was to simply secure a cylinder of limestone (calcium carbonate, $CaCO_3$) in the top of a pipe clamped in a vise, then convert it into lime (calcium oxide, CaO) using a blowtorch, and make it incandesce under a hotter oxyacetylene torch. As expected, this produced an intense white light that few people today have experienced. I decided to build a standalone spotlight version of an oxyacetylene-fueled limelight that could throw a beam like the spotlights of old.

Peter Tabur is a mechanical engineer, project manager, and technology history buff whose next projects are a model wooden ship and some fiber optic art. He's currently between jobs and looking for opportunities.

SET UP: p.107	MAKE IT: p.108	USE IT: p.114

Sam Murphy

WHITE HOT SPOTLIGHT

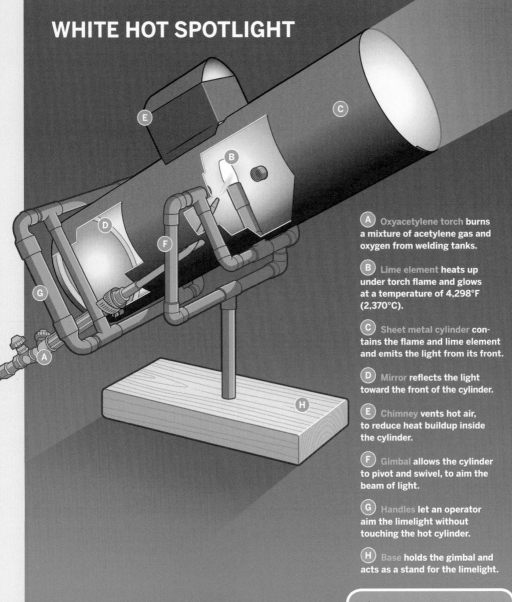

A Oxyacetylene torch **burns a mixture of acetylene gas and oxygen from welding tanks.**

B Lime element **heats up under torch flame and glows at a temperature of 4,298°F (2,370°C).**

C Sheet metal cylinder **contains the flame and lime element and emits the light from its front.**

D Mirror **reflects the light toward the front of the cylinder.**

E Chimney **vents hot air, to reduce heat buildup inside the cylinder.**

F Gimbal **allows the cylinder to pivot and swivel, to aim the beam of light.**

G Handles **let an operator aim the limelight without touching the hot cylinder.**

H Base **holds the gimbal and acts as a stand for the limelight.**

LIMESTONE, LIME, AND SLAKED LIME

Limestone is a soft, white, sedimentary stone — a precursor to marble. Like chalk, it's composed primarily of calcium carbonate. Heating limestone above 1,517°F (825°C) causes it to release carbon dioxide. This turns it into lime, aka quicklime, a caustic material used in the production of glass and steel that's also the base ingredient of lime mortar and Portland cement. Quicklime reacts exothermically with water, which is why it can burn the skin and why cement heats up as it sets.

The easiest and safest disposal method for lime is to dump it in water. This will convert it to $Ca(OH)_2$, slaked lime, aka hydrated lime.

⚠ CAUTION

• Do not touch the lime element with bare hands, as it is caustic!

• Be careful working with sheet metal, which can have sharp edges and burrs. I recommend wearing gloves, even though they make it tougher to manipulate the tools.

• Wear a welding shield or goggles when heating the lime element. It becomes extremely bright!

• The limelight generates a huge amount of heat; be careful of hot surfaces when using it, and don't leave it unattended.

Timmy Kucynda

SET UP.

A

B

C

D

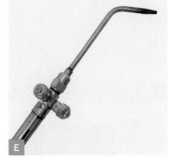
E

F

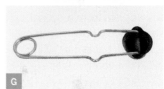
G

H

MATERIALS

A. Limestone (calcium carbonate), small piece, approximately 1"×1"×2" You may be able to get a free scrap from a local landscaping supplier, or check the garden center at any major home improvement store for limestone edging blocks.

B. Mirror, round, framed, 6" in diameter (including frame) available at most beauty supply stores. Many are 2-sided (regular and magnified) — we'll be using the regular mirror.

C. Copper pipe: ½"×6', ¾"×1'

» Pop rivets, ⅛" diameter, 1 box
» Steel, sheet stock, 26 gauge, 24"×26" non-galvanized if possible. Galvanized steel emits zinc oxide fumes when it gets really hot, and it's harder to solder.
» Copper elbow fittings, 90°, ½" (6), ½" to ¾" (4)
» Copper tee fittings, ½" (8)
» Copper coupler fitting, ½"
» Hose clamp, 2" diameter
» Plumbing solder
» Plumbing flux
» Lumber, 2×6 nominal size, 1' long

TOOLS

D. Pop rivet tool

E. Oxyacetylene gas torch

F. Oxygen and acetylene gases

G. Striker or butane barbecue lighter

H. Welding goggles or shield

» Yardstick or straightedge
» Tape measure or ruler
» Tinsnips
» Metal file
» Carpenter's square
» Center punch, conventional or spring-loaded Spring-loaded is easier to use.
» Hose clamps, 6" diameter (2)
» Cold chisel
» Hammer
» Bench vise
» Electric drill and drill bits: ⅛", ⅝"
» Sheet metal drill bit, aka castle bit This is a graduated conical bit that can drill many sizes of holes — pretty clever!

» Pipe cutter or hacksaw
» Propane blowtorch
» Dremel tool with steel cutoff wheel
» Disc sander or belt sander
» Steel wool or fine-grit sandpaper
» Disposable gloves or flux brush
» Leather or protective gloves
» Vise-grip pliers (optional)

Gregory Hayes

MAKE IT.

BUILD YOUR LIMELIGHT

Time: A Weekend
Complexity: Moderate

1. CONSTRUCT THE MAIN CYLINDER

1a. Remove any legs or stand from the 6" mirror.

1b. Draw a 20"×20" square on the sheet metal and cut it out.

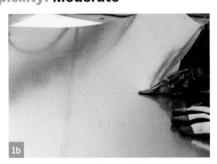

1c. Open two 6" hose clamps to their maximum diameter, then roll one long edge of the sheet metal over to the opposite edge to form a cylindrical shape. Slip a hose clamp over each end, being careful to avoid the sharp edges.

1d. Insert the mirror flush into one end (with either side facing in), and slide both clamps down to the mirror end. Cinch the 2 hose clamps to constrict the sheet metal snugly around the mirror, tightening them the same amount so that you get a cylinder, not a cone.

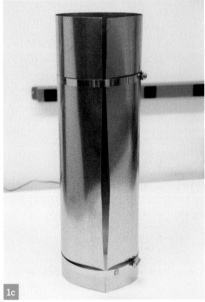

1e. Use a center punch to mark the location of the first rivet hole, where the metal overlaps near the end of the cylinder. Drill through using a ⅛" bit, then loosen the hose clamps and file away any burrs. Retighten the clamps until the holes line up again.

NOTE: Use a carpenter's square to ensure that all corners are true 90° angles. File the edges to remove burrs, and save the leftover sheet metal for later.

◤ TIP: To help hold the cylinder in place for clamping, you can secure the far end using vise-grips while you apply the hose clamps around the mirror end. If you cut an accurate square, the overlapping edges should align.

Peter Tabur

1f. Insert a pop rivet into the pop rivet tool, and insert the rivet into the hole. Squeeze the tool until the rivet "pops" (this expands the rivet on the inside until the steel pin breaks).

1g. Push out the mirror and insert it in the opposite end of the cylinder with its flat, non-magnifying side facing in. Repeat Steps 1d–1f at that end.

◤ **TIP:** To leave room for the lime holder tube, don't place a rivet exactly in the middle (i.e., 10" from either end) — you can offset it by 1".

1h. Loosen the hose clamps, slide them down the cylinder, and retighten to add more rivets spaced approximately equidistant along the length of the cylinder seam; I used a total of 5.

2. ADD THE CHIMNEY

2a. Print the chimney template from makeprojects.com/v/27 and draw the pattern onto the leftover sheet metal.

2b. Cut out the metal along the solid lines, and file to remove sharp edges. Fold as indicated on the template along the dashed lines; you can do this by bending the metal over the edge of a workbench. Drill and rivet as indicated to complete the chimney structure.

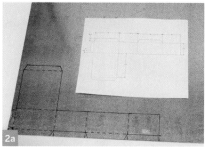

2c. Mark the center of the cylinder 180° opposite the seam. Draw a 4"×3" rectangle centered around this point, running lengthwise along the cylinder. Cut the rectangle out with a Dremel tool.

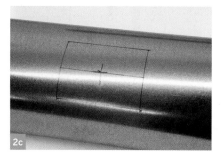

2d. Fit the chimney bottom over the hole, bending the chimney or rectangle edges as necessary so the chimney "pinch fits" snugly in place. You don't want the chimney to fall off, but you should be able to squeeze and lift it away to access the inside of the cylinder. When the proper fit is achieved, file away any burrs.

2e. Mark 3 locations at the exact middle of the cylinder's length, one in line with the rivets on the seam (the bottom of the cylinder), and the other two 90° around from this point in both directions. Use the center punch to dimple the steel at these 3 points.

2f. Drill a ¾" hole with the sheet metal bit at each of these 3 points, to just fit a copper tee fitting.

3. ADD THE GIMBAL AND HANDLES

3a. Cut the copper pipe to the following lengths and keep the leftover pieces.
½" pipe: 1⅛" (2), 1½" (4), 2" (1), 2½" (4), 4½" (4), 5¾" (4)
¾" pipe: 4½" (2)

3b. Dry-fit the pipe pieces together into a handle-and-gimbal assembly to ensure they fit together. Note that the two ¾" pipes are side-handle parts of the handle assembly, which connect via ½" to ¾" elbow fittings.

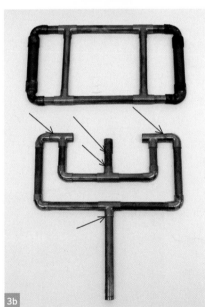

🔧 **TIPS:** Pipe cutters are relatively cheap and easy to use, and they make square cuts that are almost burr-free.

To use a pipe cutter, open it up to fit around the pipe, then tighten its handle until the cutting wheel just touches the pipe at the distance you wish to cut. Tighten the handle down another half turn (for copper pipe), roll the cutter around the pipe 2–3 times, and repeat tightening and turning until the pipe is cut. If you're cutting steel pipe, only tighten the handle ¼ turn at a time.

You can cut the pipe with either a hacksaw or a pipe cutter, but I recommend a pipe cutter. If you use a hacksaw, file each cut perimeter afterward to remove burrs.

NOTE: The 5 joints indicated by arrows will not be soldered.

COPPER PIPE SOLDERING

Soldering water pipe is an easy and useful skill to learn. After cutting the pipe to length, clean all joint surfaces with steel wool or sandpaper until they're free of corrosion, markings, etc., and have a bright shine. I wrap steel wool around the end of the pipe, then grip firmly and rotate the pipe. For inside surfaces, I wrap steel wool around a fingertip, then press and rotate the pipe around it. If you have a large project, you can buy a special steel brush ($3) that quickly cleans inside surfaces.

Once the pipe is clean, apply a thin coat of flux paste on all surfaces to be soldered. This is a chemical cleaner that's essential for sound solder joints. Apply it with a brush or wear disposable rubber gloves — it's acidic and a real pain to wash off your hands.

Assemble the parts and fire up the propane blowtorch. Keep the torch head about 3"–4" back from the pipe surface with the end of the flame in contact with the metal. Slowly move the flame around each side of the joint, and poke the joint periodically with solder. When the parts are sufficiently hot, the solder will readily melt and be wicked into the joint via capillary action.

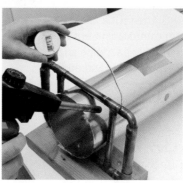

Keep applying solder until you can see a meniscus of solder around the joint's circumference. If the solder beads up rather than wicking into the joint, this indicates that the joint is not clean enough or doesn't contain enough flux. Finally, it's a very good practice to wipe away the flux when it's all cool.

3c. Disassemble the handle assembly and clean all the pieces for soldering (see sidebar, above). Pop the mirror out of the cylinder end while soldering, flux the pipe pieces, and reassemble the handle assembly around the cylinder, about 1" from the mirror end and with its sides along the cylinder's sides (seam on the bottom).

3d. Solder the handle assembly together in place around the cylinder, and solder the cylinder to the copper piping wherever they touch.

3e. Disassemble and clean the gimbal assembly pieces. As with the handles, flux the joints, and assemble the inner gimbal pieces around the cylinder, with the tee fittings running through the holes. Solder the joints.

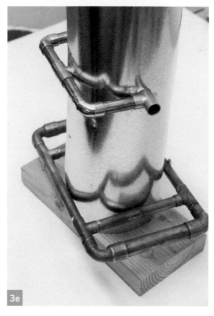

▲ TIP: I simply soldered to the sheet metal as-is, but for stronger connections, you can sand away the steel's outer galvanized layer at each solder point.

NOTE: The joints indicated by arrows in the Step 3b photo must not be soldered. These dry-fit joints will allow the gimbal to hinge and the base and incandescent element holder to be detached.

3f. Assemble the outer gimbal around the inner one, then solder all joints except those needed for gimbal rotation.

3g. To make the stand, drill a ⅝" diameter hole centered in the 2×6. Then cut a base pole of any desired length out of leftover ½" copper pipe and fit it between the wood and the gimbal's center tee fitting.

4. MAKE THE LIME ELEMENT

4a. Chisel away a chunk of limestone about the size of a walnut, then sand it with a disk grinder or belt sander into a tapered cylinder that fits into the pipe coupler.

4b. Slide the coupler with the limestone onto the 2" piece of ½" pipe. Secure the pipe in a vise, and heat the stone with a blowtorch until it glows dull red (like charcoal briquettes) for a few minutes. Move the flame around to heat uniformly and avoid cracking.

NOTE: Heating your limestone will thermally decompose the outer layer of the calcium carbonate into lime while releasing carbon dioxide.

4c. The lime element is ready. You can let it cool and install it into your limelight, but it's more fun to test it by making it incandesce right now!

Consult your oxyacetylene instructions for regulator adjustments (I used 3psi for the acetylene and 15psi for the oxygen). Start the torch and adjust for a neutral flame. Put on your welding goggles and heat the lime under the flame. Within a minute, it will become a bright white light.

5. INSTALL THE TORCH, LIME, AND MIRROR

5a. The oxyacetylene torch runs forward horizontally along the cylinder. Drill a hole for the torch nozzle to get its tip close enough to the lime element to heat it.

This was the trickiest part of the project, requiring a bit of trial and error. I cut a hole halfway between the lower gimbal hole and one of its side holes.

5b. Cut 2 slots farther back in the cylinder for threading the 2" hose clamp through vertically. This will hold the torch handle in place.

Then thread in the small hose clamp, poke the tip of the torch into the slot, and affix the torch handle to the cylinder with the clamp.

5c. Allow the stone and mounting to fully cool, then put on gloves and install it in your limelight. The 2" pipe base fits into the inner gimbal tee fitting along the seam of the cylinder.

5d. Remount the mirror. If its plastic frame can't take the heat, cut a backing plate from scrap sheet metal with 4 tabs to hold the mirror and 2 big tabs to rivet to the cylinder. Or just cut and fold tabs from the cylinder itself.

5a

5a

TIP: I found igniting the gas with the striker a little difficult in the confined space. Using a butane barbecue lighter is easier.

5b

5c

5d

5d

5d

USE IT.

LIGHT UP
THE STAGE

Limelight Performance

To use the limelight, remove its chimney, ignite the torch, and reinstall the chimney. Soon, voilà — limelight! Perfect for any non-electric night performance or steampunk cabaret.

If the lime element doesn't glow white-hot all around, grind it down smaller and move the flame close enough to envelop it. Also, try different oxygen/acetylene pressure mixes.

When I first used my limelight, I expected it to be blindingly bright from a distance, like a klieg light. I was surprised at how "not bright" it was. So I did a bit of experimenting with a photographic light meter to compare its output to other light sources.

I set the light meter at 200ASA and put it 7' in front of the limelight. I turned off all the other lights in my shop and took readings of the limelight, a 100W incandescent bulb inside the limelight cylinder, and a 750W halogen worksite lamp by itself.

For a shutter speed of 1/30 second, the light meter indicated apertures (f-stops) of 6.2 for the limelight, 5.1 for the 100W bulb, and 11.0 for the halogen lamp. This means that the limelight is a little brighter than a 100W bulb, but significantly less bright than the 750W shop light.

Coincidentally, we recently lost power to our house while cooking dinner. We lit candles and kept going. It was surprisingly dark! I had to hold a candle directly over the cookbook to have sufficient light to read. I realized that when the limelight was invented nearly 200 years ago, small flames were just about the only source of lighting. In comparison, the limelight must have seemed brilliant! ⊿

Sam Murphy

1+2+3 Action Root Beer Pong

By Cy Tymony

You can make it!

PLAYING THE POPULAR BEER PONG GAME can be enjoyable for a while, but as your skill improves, the challenge and fun diminish. After some practice, tossing a ping-pong ball at a cup just isn't too difficult.

Bring back the fun by adding motion, with your own pong toy that scuttles about.

YOU WILL NEED

Styrofoam or paper cups
Ping-pong balls
Toy motor and paper clip, or micro-vibration motor
  RadioShack #273-107, radioshack.com
Paper clips, jumbo (3)
Button cell battery, 3V RadioShack #23-804
Copper wire
Tape
Pliers

1. Make the legs.
Bend 2 paper clips into curved "C" shapes, with legs, and tape them to the sides of a cup.

2. Connect the motor.
Using a small motor removed from a toy, bend and press a paper clip around the gear shaft to make it off-balance. (Alternatively, you can use a ready-made mini vibrating motor found in an old pager or cellphone.) Test the motor with a 3V button cell battery by pressing its wires onto both terminals.

Tape the motor and battery to the bottom of the cup. Tape a short length of copper wire to one side of the battery. Tape one wire connector from the motor to the other side of the battery. Tape the 2 loose wires (from the motor and the battery) to the side of the cup so they can be twisted together, acting as a switch.

3. Play ball!
Turn the cup right side up, twist the 2 switch wires together, and the cup should vibrate and move on a flat surface. You can bend the cup's paper-clip feet so it will move in a desired pattern. Now try to toss balls into the moving cup from various distances.

Going Further
To add to the challenge, use a paper clip to attach another cup (or two) to the side of the Sneaky Action Pong cup. Draw score numbers on them to make it more competitive.

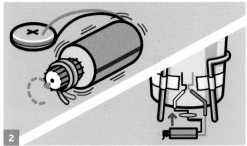

Damien Scogin

Cy Tymony is the author of the *Sneaky Uses for Everyday Things* book series. He lives in Los Angeles. sneakyuses.com.

VISUALIZING WITH IMAGEJ

Free image/video processing software creates vivid representations of time, movement, and data.

By Bob Goldstein

ImageJ is a free, cross-platform program for processing images and videos. Its author, Wayne Rasband, originally wrote it for use by biomedical researchers working with microscope images. But he designed it with an open architecture so anyone could write plugins to add new tricks to the program. As a result, its capabilities are constantly growing and improving, thanks to contributions from the more programming-savvy among its users.

But ImageJ isn't yet well-known outside of the scientific community. As a scientist and fan of creative tinkering, I thought it would be fun to introduce MAKE's readers to some of the tricks it can perform.

ImageJ processes images using filters similar to those in programs like Photoshop. These filters are written as plugins, and there are hundreds available.

In his Country Scientist column, Forrest Mims has described using ImageJ as a tool to make scientific measurements (*see MAKE Volume 18, page 42, and subsequent columns*). But my favorite tricks are the ones that transform a video into a single image, and those that perform math on images — combining two images by adding their pixel intensities, for example.

This article explains how to get started with ImageJ, then describes some of my favorite ImageJ recipes. Starting with a few interesting images or a short video captured by a digital camera or a webcam, you can cook up any of these example images within about 20 minutes.

Bob Goldstein is an occasional contributor to MAKE and a cell biologist at the University of North Carolina at Chapel Hill.

Gregory Hayes

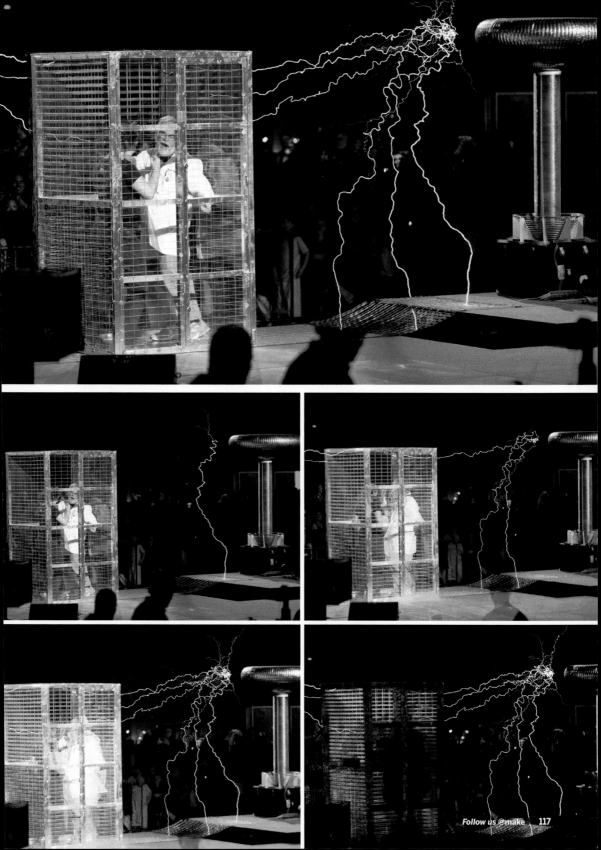

DOWNLOAD IMAGEJ AND PLUGINS

On a Mac or Linux computer, download ImageJ from the Research Services Branch of the National Institutes of Health (NIH) website, imagej.nih.gov/ij. (PC users can skip this step; the PC download has ImageJ included.)

ImageJ comes with several valuable plugins already populating its plugins folder, but you'll want to add more. The NIH website has a huge number to choose from, with bewildering names like Lipschitz Filter, so I recommend starting with the small, useful bundle published by the microscopy group at McMaster University.

Download the bundle from macbiophotonics. ca/downloads.htm and move it into the folder *ImageJ/plugins* (for PC users, this download includes the ImageJ application, so the one download is all you'll need).

The McMaster site includes some simple instructions for allocating system memory before you begin. ImageJ is a memory hog, and following these directions before you launch the program will save you headaches later.

Once you've installed ImageJ, it can be interesting to open a photo or import a video and just start clicking away to see what various buttons do. For a more systematic start, you can follow the User Guide, available under the Documentation link at rsbweb.nih.gov/ij. The Plugins link on the same page lists additional plugins with short descriptions, and you can install most of them by simply moving them into your plugins folder.

CREATE IMAGES

Here are some examples of what you can do with ImageJ. For each, I'll give a short description, sample images, and a step-by-step protocol that explains how to do it with your own photos and videos. The letters (A–F) correspond to the images shown.

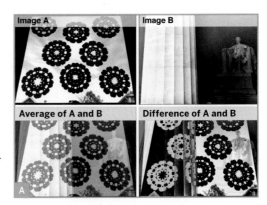

Image A | Image B
Average of A and B | Difference of A and B

A. IMAGE CALCULATOR: TWO IMAGES COMBINED

The Image Calculator tool can combine images using simple arithmetic; for example, by adding, subtracting, or averaging the colors of each pixel. In the 4 sample images above, the lower left-hand corner pixel of the snowflake image has an RGB color of 12,6,10, while the lower left Lincoln pixel is 70,54,42. (RGB encodes colors as 0–255 each for red/green/blue, where 0 is darkest and 255 is brightest.) Averaging these two values results in a pixel with a color of 41,30,26. Image Calculator combines the pixels of all images this way.

Scientists might not need to combine images of paper snowflakes and the Lincoln Memorial, but the Image Calculator can be a powerful tool for combining other images in predictable ways. For example, it brings out differences between 2 nearly identical photos, as in Garry McLeod's photos (A), opposite.

Protocol: Click File → Open to select an image on your computer, and repeat to open a second image. Click Process → Image Calculator, select each of your image names, then select an operation like Average or Difference to combine the images.

B. Z-PROJECT A STACK: LIGHTNING FLASH

Here I started with a short video I captured of a lightning flash, which took about a second to cross the sky. On the video, distinct parts of the flash could be seen in separate frames (Figure B, frames 1 through 8). To see

Bob Goldstein

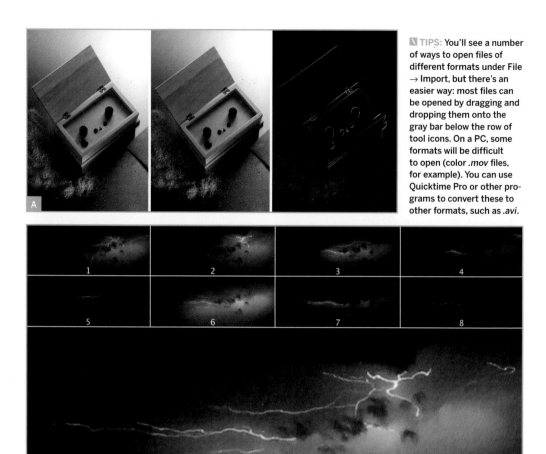

what the entire lightning bolt looked like, I combined all the video frames into one image using the Z Project tool. This effectonly works well with video captured from a stationary camera position.

Protocol: Before opening a video file, you may need to use a video editing program to shorten its length or lower its resolution. ImageJ has no problem with 800×600 pixel videos of a few hundred frames, but longer videos with much larger dimensions can slow or crash the program.

Once your video is opened in ImageJ, click to select it. (You can scroll through it using the < and > keys.) Click Image → Stacks → Z Project, and choose Max Intensity to add up the brightest pixels from each frame.

(The montage in Figure B was also made with an ImageJ tool. Select your video and

click Image → Stacks → Make Montage to see options for making montages.)

C. Z-PROJECT A STACK: PELICAN FLIGHT

The image in Figure C was made in a similar way to the lightning flash, but here the subject was darker than its background. The video was taken by a stationary camera, pointing at the

Garry McLeod (A)

sky as birds flew overhead. I then Z-Projected a 2-second segment, about 30 frames.

Protocol: Open your video. Click on your video to select it, then click Image → Stacks → Z Project, and choose Min Intensity to add up the darkest pixels from each frame.

D. COLOR-CODING TIME: STAR PATHS ACROSS THE NIGHT SKY

I made a time-lapse film of the stars passing over my yard from dusk to dawn. (I used a Canon camera hacked with open source Canon Hack Development Kit, or CHDK, software to do long-exposure time-lapse recording).

I used ImageJ to make a black and white film of only the moving objects by subtracting everything from each frame that was also in the previous frame. I then time-coded the resulting film by color, with purple slowly turning to yellow.

Lastly, all the colored frames were projected onto a single image. In the final image, purple objects were the ones visible just after sunset, and orange/yellow objects, like the clouds, appeared around sunrise. Stars moved across the field of view at all times.

Protocol: Open a video. When the "convert to 8-bit grayscale" option pops up, accept it. Click on the video to select it, and make a second video showing only the moving objects by clicking Plugins → Stacks — T-Functions → Delta F. To color-code the time, click Plugins → Stacks – Z-Functions → Z Code Stack, and select a color scheme.

Finally, click Image → Stacks → Z Project, and choose Max Intensity to add up the brightest pixels from each frame.

E. KYMOGRAPH: MOTHER ROBIN'S NEST TIME GRAPH

My sons and I found a robin's nest on our house, and we were fascinated to have a peek. So we set up a webcam and watched the nest. One day we made an all-day, time-lapse recording. The mother sat on the eggs throughout the day, periodically taking trips away, presumably for food.

We were curious to see if there were any obvious patterns to the timing of her trips, so we made a kymograph, which displays a single slice of the image along one dimension, with time running along the other dimension.

The kymograph in Figure E marks the time in 10-minute intervals along the top, running from morning at the left to evening at the right. When the blue of the eggs is visible in the vertical stripes, the mother bird was out of the nest, and you can see how the egg positions changed with each maternal visit.

We had read online that robins never leave their nests for more than 10–15 minutes at a time, but it looks like this bird took a long lunch from around 11:40 a.m. until noon. It got dark just before 7 p.m.

Protocol: Open a video, then select the straight line tool from the row of tool icons, and click and drag to draw a line over an area of interest in the video (here, the eggs). Click Image → Stacks → Reslice to see what happened under that line over time.

F. IMAGE PROCESSING WITH A SPREADSHEET

You can extract the pixel values from an image and transform them yourself in a spreadsheet program like Microsoft Excel. It's interesting to see how features like local contrast can be highlighted this way.

Here, I started with a 50×50 pixel image of an acorn (Image 1 above), then generated several versions using simple formulas in

 +  =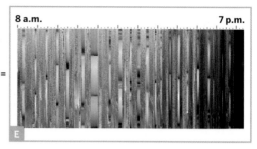

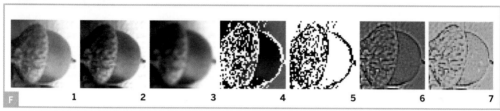

Excel. Image 2 is a grayscale version of 1. In 3, I blurred the image by averaging each pixel's value with that of its neighbors. In 4, I highlighted high-contrast areas by comparing each pixel with its neighbors. Image 5 shows only the high-contrast areas. In 6, the contrast values that I used to make 5 were converted into continuous grayscale (6), and in image 7, this grayscale image was converted to color (in ImageJ, not Excel) using a lookup table.

Protocol: First, I recommend that you select (or crop or resize down to) a tiny image of about 50×50 pixels, as each step that follows will work very slowly on larger images.

Save an image using File → Save As→ Text Image. This produces a text file that you can open in a spreadsheet program. In the spreadsheet (I used Excel), you'll see an array of numbers representing the pixel values in a grayscale version of your image. In Excel, I transformed the pixel values using a variety of formulas (see makezine.com/27/primer for a sample spreadsheet).

After manipulating the numbers in the spreadsheet, I copy-pasted those cells into a text editor (I used TextWrangler for Mac) and saved it out as a plain text file (.txt). Back in ImageJ, I opened the file with File → Import → Text Image. For the final transformation, from Image 6 to Image 7, I clicked Image → Lookup Tables and selected a color scheme for coloring the grayscale image.

CONTRIBUTE TO SCIENCE!

Makers and scientists both constitute creative communities that could learn a thing or two from each other. If you can program in Java, and you see an interesting way to display images that no existing ImageJ plugins can yet do, why not write a new plug-in? Hundreds of plugins exist already, but the best ones probably have yet to be written.

If you do write a new ImageJ plugin, you can add it to the wiki at imagejdocu.tudor.lu.

Who'll make use of your plugin, and what scientific discoveries might it help propel? Biomedical research articles are increasingly found in full form online, so in the months and years to come after submitting a plugin, search online for its name to find out! ☑

RESOURCES

» ImageJ software: rsbweb.nih.gov/ij
» McMaster Biophotonics Facility ImageJ software library: macbiophotonics.ca/downloads.htm
» ImageJ Information and Documentation Portal — includes a wiki where new plugins can be added: imagejdocu.tudor.lu
» FIJI — a distribution of ImageJ with Java and well-organized plugins: fiji.sc.wiki

1+2+3 Stud Chair
By Corky Mork

RECENTLY, AT MY LOCAL HOME CENTER
I stopped to look at the 2×4 studs. They cost just a few dollars, and I often take a few home "just to have." They're handy for workbench legs, sawhorses, shelving, mailbox posts, temporary staging, and many other uses.

I got to thinking, *What could I make with just one 8-foot 2×4?* I got out my sketchpad and worked it through ... Yes! I could do it.

I set to work with my table saw, and in less than an hour, I had a perfectly serviceable chair! It's not the prettiest or most comfortable, but for the price in materials and effort, it's hard to beat.

1. Cut the 2×4.

Cut the 2×4 into one 32" and four 16" lengths, then rip these pieces lengthwise. The following rip measurements assume a ⅛" kerf.

Rip the 32" piece into thirds (1³⁄₃₂" thick) for the legs, then cut one of these into two 16" legs. Rip the 16" pieces into 8 slats ¹¹⁄₁₆" thick.

2. Screw and glue.

A square can help keep things, well, square during assembly. For each joint, square it up, drill pilot holes to avoid splitting, and assemble with glue and 2 drywall screws. The glue will help keep the chair from racking (twisting out of square), especially on the leg joints.

First build the 2 side frames; one is mirrored from the other. Add slats to the front and the seat back, then add the seat slats.

3. Finish it, or not.

You can paint, stain, and decorate your chair any way you like. I kept my first one *au naturel* and left it outdoors to develop a rustic look.

What can *you* make with an 8-foot 2×4? Maybe a better chair? A table? Or what? ◪

Corky Mork has been making stuff from wood, electronics, toys, and junk since he was a kid. When not making stuff, he acts in community theater, sings a capella, and plays the tuba.

YOU WILL NEED

Dimensional lumber, 2×4, 8' length straight and smooth, with as few knots as possible. A nominal 2×4 actually measures 1½"×3½".
Drywall screws, 1⅝" (32)
Wood glue
Saw I recommend a table saw or band saw, though a portable circular saw or even a handsaw could work if you're careful.
Drill with ⅛" twist bit and Phillips screwdriver bits
Square (optional)

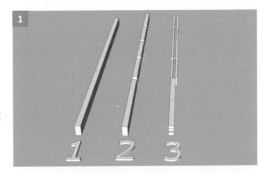

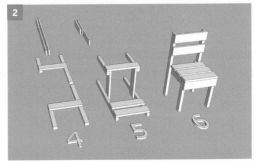

Illustration and photo by Corky Mork

Spoon-Carving Knife
Make a blade to make useful things for the kitchen.
By Doug Stowe

Doug Stowe

IN RECENT GENERATIONS, WE'VE LOST many basic skills. Today, making a simple knife to work wood is beyond the capacity of most people in the Western world. Sad, but true. And yet it's fun, and offers a sense of tactile satisfaction. The use of this knife is therapeutic, leading to feelings of accomplishment.

And knives are tools. Sadly, most have come to regard knives as weapons, forgetting the simple pleasures of whittling on a stick or making something useful for the kitchen.

Making a Simple Spoon Knife
You can't buy anything as effective as this knife in a store, so you'll need to make your own. It's designed for carving the bowl shape at the working end of a spoon. With this knife and a sloyd (whittling) knife, you can carve anything you want. In fact, you can remake

yourself into a craftsman. It just takes a little carving. But, I'll remind you that you won't find everything you need in this article. It takes practice to actually get good at something. Your first efforts may not result in what you see here, but things are made more meaningful and useful by the effort we've invested in them.

1. Grind the blade.
Use a grinder to shape the steel blank and begin forming the knife edge (Figure A, following page). For a right-handed knife, hold the steel from the right side of the wheel. For a left-handed knife, work from the other side.

2. Bend the blade to shape.
Use pliers to bend the end of the sharpened steel blank (Figure B). The degree of curvature

in the blade determines how deep you can go into the bowl end of the spoon. Study the spoons in your kitchen to determine the level of curvature you'll want. The blade should be curved slightly more than is required to conform to the spoon shape.

3. Drill holes for the pins.
Use a ³⁄₁₆"-diameter drill bit in the drill press to make holes for the brass pins that will secure the handle (Figure C).

4. Heat treat the blade.
Use a propane torch to heat the end of the knife blank to cherry red (Figure D). This is best done in the dark. You'll see the steel turn a variety of colors as it heats up. When it's cherry red, quench it quickly in oil. This hardens the steel so the knife will hold an edge, but it also makes the steel brittle, requiring the next step.

5. Temper the steel.
Preheat the oven to 425° and place the metal cookie sheet inside. When the oven has reached temperature, put the heat-treated knife blade on the cookie sheet for 20 minutes. Then remove it and allow it to cool to room temperature.

6. Make scale stock.
Rip wood stock as shown in Figure E to form the 2 "scales" that make up the handle of the knife. A ripping cut in one scale, only partially through, provides a secure housing for the knife blank on 3 sides as shown.

Cut the scales to length to fit the blade.

7. Glue scales on both sides.
Use epoxy to glue the knife blank into the grooved scale (Figure F, following page). Then use the drill press to finish the holes through the scale.

Glue on the other scale using epoxy, and then drill to finish the holes through to both sides. Whether you're making a left- or right-handed knife, yours should look like one these (Figure G).

MATERIALS
Steel blank, approximately ³⁄₃₂"×½"×5" for blade
Wood stock, walnut: ³⁄₈"×1"×18" and ³⁄₁₆"×1"×18"
Brass rod, ³⁄₁₆" diameter, a few inches long You'll cut it down to 2 pins ⅞" long.
Epoxy glue
Sandpaper, 600 grit, extra fine
Danish oil, ½ pint

TOOLS
Grinder	**Cookie sheet, metal**
Pliers	**Clamps**
Drill press (preferred)	**Sander**
or electric drill	**Hacksaw**
Propane torch	**Dowel**
Oven	

⚠ **CAUTION:** It's important to use a fence on the drill press. If the drill were to catch and the blade to spin accidentally, you could be cut.

A

B

C

D

E

8. Shape the handle and install the brass pin.

Use a sander to shape the edges of your knife to be comfortable in your hand (Figure H). Of course another carving knife could be used for this! But either way, this is a job that requires skill and attention.

Cut ³⁄₁₆" brass rod into 2 pins ⅞" long and glue them in the holes using epoxy glue.

9. Finish and sharpen.

After sanding (and sanding and sanding) apply a hand-rubbed Danish oil finish to the knife to protect the wood and bring out its beauty (Figure I).

Use a dowel wrapped in 600-grit extra-fine sandpaper to sharpen the cutting edge (Figure J). Use a flat sheet or sharpening stone to hone the backside of the blade.

10. Carve a spoon.

Now you're ready to carve a spoon. You'll also need a sloyd knife or other type of whittling knife as shown in Figure K.

You'll find that green wood works best for carving spoons. So cut a stick, use an axe to split it in two, and then begin carving (Figure L). The spoon in Figure K is made of basswood. Many other woods will work as well.

Also, if you can make a spoon knife, there is no reason you couldn't make a sloyd knife as well. The technology is the same except that traditional sloyd knives, like this one from Mora, Sweden (Figure K), are made with many laminations of steel — a technique hard to duplicate outside the blacksmith shop. ◪

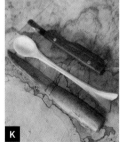

Doug Stowe lives in Arkansas' Ozark Mountains, where he makes, teaches making, and writes about both. He's working on his seventh woodworking book and is best known for box making.

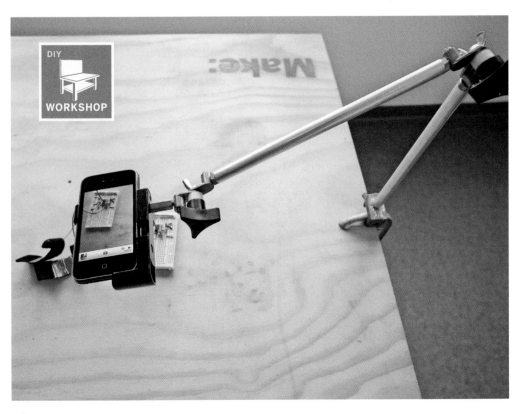

$30 Gobo Arm: Mobile Document Camera Stand

Go hands-free for the price of a clamp.

By Adam Flaherty

IF YOU'RE FAMILIAR WITH TABLETOP product photography and video, you know what a pain it can be to shoot overhead shots while working on your subject. Tripods just get in the way.

The usual remedy is a document camera, the modern equivalent of the overhead projector. These cameras let you display close-up shots with an LCD projector or external monitor, and they cost between $500 and $1,500. They're well suited to capturing images of document-sized objects, but come on, a grand for a glorified webcam?

We've all got streaming video on our smartphones now, so couldn't we just use that instead? The problem is, how to hold it over

the project without a tripod to bump into.

Hollywood grips solve this problem with what is known as a gobo arm, a lightweight mount that lets you position your smartphone (webcam, etc.) down where the action is without getting in the way. Commercial units run $100–$150, not including the $30 clamp you attach it to. But for the price of one of those fancy clamps, you can piece together your own Mobile Document Camera Stand using easily sourced parts.

It's easy. If you can make cuts with a hacksaw, you can build this project. There's also plenty of room for improvement. For example, I recently upgraded the lock washers to the "curved disc spring" type, and they work great.

Sam Murphy

1. Cut rod and tube, and assemble.

Cut two 11½" lengths and one 1¾" length of threaded rod. Cut two 10" lengths of aluminum round tube. Clean up the ends with a file.

Insert the threaded rod into the aluminum round tube, and use hex nuts to hold in place. Optionally, you can use nylon spacers or 8" strips of electrical tape wound around the threaded rod toward each end to position the tube. You should have enough rod left at each end to fit into the terminal lug.

Attach a single terminal lug to one end of the first threaded rod, and tighten with a slotted screwdriver. Attach terminal lugs to both ends of the second threaded rod, and tighten. Then thread the coupling nut onto the 1¾" threaded rod. Attach the remaining terminal lug to the end and tighten (Figure A).

2. Assemble the friction joints.

For the first friction joint, insert a nylon spacer onto the clamping knob shank. Insert the end of the terminal lug over the nylon spacer, keeping the locking screw facing away from the tip of the shank. Insert a lock washer over the terminal lug.

Now, insert the end of the terminal lug of the second threaded rod over the lock washer, keeping the locking screw facing toward the tip of the shank. Hold it together with a ¼"-20 wing nut (Figure B).

Repeat for the second friction joint, to attach the 1¾" threaded rod assembly to the other terminal lug of the second 11½" threaded rod assembly.

3. Finish.

Thread the beam clamp onto the remaining end of the first threaded rod assembly and attach it to the C-clamp (Figure C). Make sure a little extra of the threaded rod sticks through the bottom of the beam clamp. You'll want this overhang to wedge the C-clamp into so that the beam clamp aligns straight.

Attach the generic smartphone tripod adapter to the ¼"-20 end of the 1¾" threaded rod assembly. Use the coupling nut to adjust for a snug fit.

MATERIALS AND TOOLS

See makeprojects.com/v/27 for recommended suppliers, prices, and other sourcing information.

Threaded rods, ¼"-20, 12" long (3)
Setscrew terminal lugs, ¼", straight tongue, copper (4)
Spacer, ½" OD, .252" ID, ½", nylon (2)
Wing nuts, ¼"-20 (2)
Aluminum tubing, ½" OD, 0.4" ID, 10" length (2)
Clamping knob, threaded stud, 3-arm, ¼"-20, plastic (2)
Lock washer, ¼", curved disc spring (2)
Coupling nut, ¼"-20
Beam clamp, ¼"-20
C-clamp
Smartphone camera adapter, generic Check eBay or theglif.com (for the fancy iPhone 4 model).
File for cleaning up cut rod
Screwdriver, slotted
Hacksaw

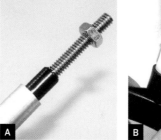

A

B

C

Now you're ready to photograph your project for sharing at makeprojects.com, or even stream live video from your workbench. ◪

Adam Flaherty is a contributing writer for makezine.com. He edited MAKE's *iPhone Hacks* book (O'Reilly), and has tech-reviewed other O'Reilly titles including *Linux Unwired* and *Netbooks: The Missing Manual*. He has dabbled in industries including alcohol, tobacco, firearms, and the declining automotive industry.

Adam Flaherty

Portable Powerhouse

This semi-flexible solar panel system clips onto a backpack and provides power to electronic devices.

By Andrew Lewis

WHEN THE TIME CAME FOR ME TO SELL my cabin cruiser, I couldn't bring myself to part with the 17-watt solar panel that I'd fitted to its deck. This panel was my first foray into renewable energy, and it had served me well aboard my 22-foot fiberglass home away from home.

I turned my attentions to camping, and found that the problem with getting back to nature is a lack of charging facilities for the computer, MP3 player, and cellphone that I like to use. I tried some portable solar chargers but found them either too low-power or too awkward to comfortably carry.

I soon realized that the semi-flexible, resin-coated solar panel from my old boat was roughly the same size as my backpack,

so I set about repurposing this solar trophy to suit my needs. For strapping the panel over the backpack, I attached fabric storage pockets to its top and bottom, and I outfitted the bottom pocket with a flat "connecting plate" that sports 3 pairs of metal snaps, spaced apart to prevent electrical shorts.

The connecting plate attaches to custom charging cords with a snap pair at one end and a USB plug on the other. By plugging the cords into the plate, you can trickle-charge up to 3 USB devices at once in bright sun.

1. Make the connecting plate.
Detach the solar panel's power cord (Figure A). I removed a plastic cover, then cut away some silicone from the wire leads underneath

MATERIALS

Solar panel, semi-flexible, no wider than about 12" Choose a size that will fit over the top and back of your favorite camping backpack without hanging over the side and getting snagged on the scenery.

Fabric, waterproof, 1 yard square Remember that waterproof and water resistant are not the same.

Plastic sheet, thin but rigid, about 6"×4"

Metal snaps (8) Any size, but they must be conductive metal; test with a multimeter if you're in doubt.

Copper or aluminum tape, 1" wide, 1 roll Copper is better, but aluminum will work fine.

Wire, 18-gauge, insulated

Spade connectors, medium (optional) to fit insulated wire

Diode, power, 1N4002 or similar (1) plus more for each charging cable; see below.

Acrylic PCB sealant, spray can

Bolts, stainless steel, M4×10mm long or similar size (4)

Nuts and washers to fit bolts (4) Ordinary nuts will be OK, but stainless will resist the weather longer.

Velcro tape, 8"

Nylon strapping, 1"×12" A luggage strap is ideal.

Thread for sewing

D rings, metal (4) the type used on luggage or saddlery, to fit the strapping. Or you can use buckles.

Heat-shrink tubing, assorted sizes

Tape, double-sided

Masking tape or chalk

White (PVA) glue (optional)

For each charging cord (charger can accommodate up to 3 cords at once):

Diodes, power, 1N4002 or similar (4)

Cable to fit chargeable device: USB cable for USB devices, and laptop cable for laptop.

Voltage regulator The voltage out should match the device. I used a 78M05 (5V) for my USB cable and a 78M12 (12V) for my laptop cable.

TOOLS

Hammer

Hole punch sized for attaching snaps

Sewing machine, or needle and thread

Soldering iron

Wire cutters

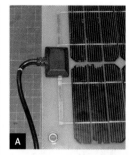

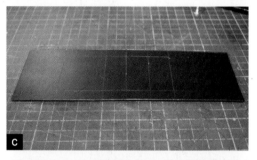

(Figure B) and desoldered them.

Punch 2 rows of 4 small holes at regularly spaced intervals lengthwise along the 6"×4" plastic sheet (Figures C and D). Space the rows of holes fairly far apart; they'll accept the backing pieces of the snaps carrying power from the solar panel, so they shouldn't be easily bridged. Wrap 2 strips of copper or aluminum tape all the way around the sheet and over each hole row (Figure E). This will make the electrical connections between the snaps.

Strip both ends of two 7" lengths of wire. Cut one wire in the middle, and solder-splice in a 1N4002 (or similar) diode. This diode will allow power to flow away from, but not back into the solar panel. Mark the wire on the marked (cathode) side of the diode, and then insulate the joints with heat-shrink tubing.

Push 2 snap backing pieces through the pair of holes at one end of the connecting plate. Wrap the cathode end of the diode wire and either end of the other wire around each backing piece. (If the wires are too thick to wrap comfortably, solder a spade connector onto each and insulate the joint with more heat-shrink.) Hammer a male top piece for each metal snap into place using the hole punch, making sure that the wire ends or spade connectors are sandwiched securely between the backing and the tape (Figure F).

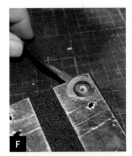

2. Make the electrical and storage pockets.

The plate is shrouded at either end by a waterproof pocket, similar in design to a pencil case. One pocket contains the electrical connections to the solar panel, and the other can be used to store cables or batteries.

Cut a piece of waterproof fabric a little wider than the solar panel and about 10" long. Position the connecting plate with its widest edge about 2" away from the edge of the fabric, then mark and cut small holes in the fabric where it lines up with the 6 holes in the plastic. Poke the backing pieces of the 6 unmounted snaps up through the plastic plate and the waterproof fabric, then hammer the snap tops to attach them over the fabric (Figures G and H).

Spray the back of the connecting plate (and any other exposed areas) with acrylic PCB sealant. The idea is to waterproof as many of the exposed electrical connections as possible, while leaving the snaps uninsulated.

Fold the fabric in half underneath the plate so that the snaps face upward, and stitch the metal D rings to the folded edge using short loops of nylon strapping (Figure I, bottom).

Test-fit the open long side of the pocket against an end of the solar panel. Stitch the fabric pocket along its shortest sides, positioning the stitching runs so that the pocket fits tightly over the end of the solar panel.

Now slip the pocket over the solar panel so that the panel doesn't quite meet the connecting plate inside. Using chalk or tape, mark

a seam line on the pocket, midway between the panel and plate. Also mark the locations of the solar panel's corner mounting holes.

Remove the solar panel from the pocket, and stitch in from the sides along the seam mark you just made. But don't connect the seam all the way through; leave the wires from the connecting plate poking outside the pocket. You should have a closed bag with 2 wires sticking out from the side (Figure I, top).

Solder the wires from the plate onto the solar panel, with the diode wired to the positive (+) terminal. Poke any excess wire back inside the fabric pocket.

Make holes in the pocket at the mounting locations you marked, and attach the solar panel to the pocket via the mounting holes, using nuts, bolts, and washers. If you want to reinforce the area where the bolts pass through the material, I recommend painting the edges of the holes with white glue.

Spray the solder joints on the solar panel with the PCB sealant, and stick the loose fabric seams to the solar panel with double-sided tape.

Make a storage pocket from waterproof fabric, and attach it to the other end of the solar panel, just like you did with the electrical pocket (Figure J). The only differences are that the storage pocket will be empty, and should be held closed with velcro instead of with stitches (Figure K). To make sure the velcro lines up correctly, remember to sew it in place before you stitch the other seams closed.

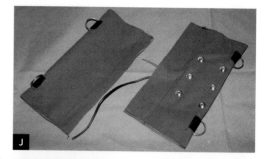

3. Make the charging cords.

For each charging cord, take 2 strips of waterproof fabric, and add snap tops (female) to one piece, spacing them apart the same distance as the snaps on the connecting plate. Solder wires directly onto the backs of the 2 snap tops and spray them with PCB sealant.

I used a bridge rectifier circuit to control the polarity of the power from the snaps, so it wouldn't matter which way I snapped on the cord.

Make this circuit by soldering 2 power diodes oriented in opposite directions off of each snap wire, 4 diodes total. Solder the diodes' 2 unconnected cathode ends together to the input pin of the voltage regulator (pin 1), and the free anode ends of the other 2 diodes to the regulator's ground pin (pin 2). Insulate all connections with heat-shrink.

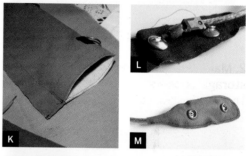

Cut, split, and strip the power supply end of your device's charging cord and identify the power and ground wires inside. With a USB, the power wire will connect to pin 1 and should be red; the ground wire will connect to the last pin, 4 or 5, and should be black. Solder the regulator's ground pin to the charging cord ground, and its output pin (pin 3) to the charging cord's power wire. Insulate with more heat-shrink (Figure L).

Finally, stitch the other piece of fabric over the wires and snaps to enclose the connections (Figure M). Use a piece of heat-shrink

tubing to reinforce the stitches where the cord exits the cloth sleeve. The cord is done, and you're ready to plug into the sun!

The flexible pockets and D-rings at the ends of my Portable Powerhouse let me strap it to my ordinary book bag as well as my camping rucksack, so I can charge my laptop while I'm walking around town (Figure N).

My next project will be to add a lithium polymer (LiPo) battery pack with an integrated charge circuit. This will enable the Portable Powerhouse to store power for use at night, after the sun goes down! ✂

Andrew Lewis is a keen artificer and computer scientist with interests in 3D scanning, computational theory, algorithmics, and electronics. He is a relentless tinkerer, whose love of science and technology is second only to his love of all things steampunk.

Fishing for Swarms

Lure in a local colony of honeybees with a simple wooden bait hive.

By Abe Connally and Josie Moores

GETTING HONEY FROM YOUR "HONEY Cow" beehive (*Volume 25, page 123*) requires that you have bees in it. In this article we'll show you how to make a bait hive and some bee lure and go fishing for a swarm of bees to fill your Honey Cow.

Bees tend to swarm during the big nectar flows in late spring and early summer, so you should set up your bait hives well ahead of this time. It helps to have more than one bait hive, to increase your chances. In some areas, it's possible to catch a few swarms per year.

Like the Honey Cow, the bait hive has a series of bars arranged side-by-side along its top. The bees attach their honeycombs to the underside of the bars. These top bars are interchangeable, so it's quick and easy to move

your fresh-caught bees to the Honey Cow.

As with fishing, you must be patient. But before long, you'll have happy and productive bees ready to join your family.

1. Build the bait hive.
Our bait hive has 6 top bars, for an interior width of approximately 8½". This interior width may vary according to the width of your bars.

1a. Start by measuring the total width of the 6 top bars. If they are more than 8½" in total width, use 5 top bars instead. This measurement will affect the length of the sides and bottom. In this version, the sides and bottom pieces are 10½" long (8½" plus 2" for the thickness of the lumber).

MATERIALS

Lumber, untreated, 1×10, 8' length cedar or pine work well

Lumber, 1"×1½", 24" lengths (5) for top bars. If you've built the Honey Cow, use those bars.

Wood screws, 1½" (24)

Wood screws, ¾" (10)

White paint a small amount

Linseed oil, boiled (optional) or other natural wood preservative

Sheet metal, 11"×3" (2)

Beeswax, 1" cube

Olive oil, ¼ cup

Lemongrass oil (20 drops) 100% natural or organic

TOOLS

Electric drill, cordless

Circular saw

Drill bits, ¼" and ¾"

Wood rasp

Straightedge

Paintbrush

Fluorescent marker such as a yellow highlighter

Small glass jar and lid

Stove and small saucepan

Beekeeper gear, including a smoker, gloves, full protective suit, and veil

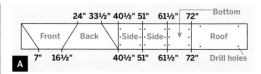

1b. Mark one edge of an 8' length of 1×10 lumber as follows: 24", 33½", 40½", 51", 61½", and 72". On the opposite edge of the board, mark the following: 7", 16½", 40½", 51", 61½", and 72". Draw lines between the marks, to create your cutting template.

1c. Using a circular saw, cut the board according to the diagram (Figure A). Also cut top bars from the 1"×1½" lumber if you haven't built the Honey Cow.

1d. On the cut edges of the side and bottom boards, drill 3 holes along each side (6 total per board), using the ¼" drill bit. Make the holes about ½" from the cut edges. On the roof board, drill 4 holes, 2 on each side on the factory (uncut) edge.

1e. On a flat surface, place the front and back boards upright with the 24" edges down. There should be an 8½" to 8¾" space between the boards. Lay the bottom board atop the front and back boards and screw in

place, carefully keeping the boards square (Figure B).

Test the width of your bait hive by placing your top bars in between the face boards. They should fit easily, but not have any spaces between the top bars. Now, using the 1½" screws, screw in each side board, carefully keeping the width of your box consistent (Figure C).

1f. Once everything is screwed together, use a ¾" drill bit to make an entrance hole on the bottom of one of the sides (Figure D). Slightly round the edges of the hole to give it a nice, funnel-like opening.

1g. You can weatherproof the outside of your hive with a natural preservative, like boiled linseed oil and beeswax. Try to avoid chemical preservatives, as they will add a smell to the hive that might repel the bees. The preservative is optional, depending on your climate (in our climate, there's really no need).

1h. Paint around the opening of the hive with just a bit of white paint. Once the paint is dry, add some lines and decorations around the opening with a yellow highlighter (Figure D, previous page). Bees see fluorescent colors very well, and having a contrasting color around the entrance will make it easier for the swarm scouts to find the front door.

1i. Place the top bars across the top of your hive (Figure E). Attach the pieces of sheet metal on each side with ¾" screws, to prevent the top bars from moving or shifting (Figure F). Then, place the roof on top.

2. Make the bee lure.

2a. Place a small glass jar in a saucepan, and raise it about 1" off the bottom of the pan on a small metal rack or something similar. Put some water in the saucepan, about halfway up the jar.

2b. Put the saucepan on the stove and slowly heat the water to a simmer (not boiling). Once the water is simmering, place ¼ cup olive oil and the 1" cube of beeswax into the glass jar, stirring constantly (Figures G and H).

2c. As soon as the wax is melted, turn off the heat. Add 15 to 20 drops of lemongrass oil to the wax/olive oil mixture and stir well.

Take the glass jar out of the water and allow it to cool. This concoction makes a great bee lure for attracting your bees.

2d. Rub a bit of the bee lure on the underside of each top bar in your bait hive, and a bit more around the entrance. You don't need much, just a dab in these key spots.

3. Set the trap.

Setting up your bait hive is the most important step. Experience shows us that you can improve your chances of catching a swarm if you do the following:

» Situate a hive in a good bee location: a place that has water and plenty of flowers, and isn't likely to be disturbed by humans.

E

F

G

🔲 **TIP:** Cut the beeswax into smaller pieces to expedite the melting process.

H

» Place the hive at a height of anywhere between 8' and 15'. Putting the hive in a tree or on a roof works well.

» The hive shouldn't be in full, direct sunlight — dappled sunlight works well.

» Face the entrance of the hive toward the sun, meaning it will face south in the Northern Hemisphere and north in the Southern Hemisphere.

» If you have a bit of old brood comb, it's good to place this in your bait hive as well.

» Make several bait hives and place them all around your area.

» Once you place the hive, leave it alone for 2 to 3 weeks. When you do check on the hive, just look at it from a distance and try not to handle it very much. You can add more bee lure about once a month.

4. Transfer the hive.

If you're lucky, one day you'll check on your hive and you'll see bees flying in and out of the entrance. Once you notice bee activity, leave the hive alone for another week and then visit it after sundown. No bees should be leaving the hive at this time of the day.

4a. Place a bit of rag or cotton in the entrance to block it so that no bees can enter or leave. Carefully transport the hive to your Honey Cow or other top bar hive. For now, just set the hive next to the Honey Cow. You'll come back tomorrow to continue the transfer.

4b. In the morning, suit up again, and start your smoker. Make room in your Honey Cow for the 6 new top bars. You should place them in the middle of the hive, at least 5 or 6 bars from the entrance. As you move the new bars in, keep them in the same order.

4c. Puff a bit of smoke toward the bait hive. Carefully remove the roof and puff a bit more smoke over the top of the top bars. Starting at one end, remove a top bar and carefully lift it straight out and quickly place it in the Honey Cow. Move the second bar the same way, carefully and quickly.

⚠ CAUTION: It's important to wear protective clothing while handling and transferring the hive.

NOTE: Blowing a bit of smoke into a beehive makes the bees lethargic by way of honey ingestion. The bees smell smoke and think there's a fire threatening their home. They then focus their efforts on eating the honey (to protect it from the fire), and they don't pay attention to you opening their hive.

4d. Once all the bars have been moved to the Honey Cow, take out 6 bars at the back of the Honey Cow. Quickly lift the bait hive and place it upside down on the Honey Cow, and then bump it quickly to drop the bees out.

4e. Leave everything alone for 30 minutes or so, to give the bees a chance to move into the new space, then remove the bait hive. If there are still a lot of bees in the bait hive, quickly hit it once or twice against the top of the Honey Cow to knock the bees down into the new hive. Using a bee brush or a leafy branch from a tree, brush the remaining bees into the Honey Cow.

4f. Replace the missing top bars in the Honey Cow and put on the roof. Put the bait hive in your house or another area where bees can't get to it. Leave the Honey Cow alone for a while, at least 2 weeks. Check on it periodically, to confirm that the bees are entering and leaving. If all is well, 2–3 weeks after the transfer, you can open the Honey Cow and check on your bees. They should be starting to enlarge any existing combs and possibly building new combs. You have successfully captured and transferred a swarm! ◢

Online Resources
» Tools, accessories, and DIY kits for top bar hives: goldstarhoneybees.com
» Natural beekeeping forum: biobees.com
» Authors' site: velacreations.com/bees.html

Abe Connally and Josie Moores are an adventurous couple living in an off-grid hideaway with their 2-year-old and newborn. Their experiments with energy, architecture, and sustainable systems are documented at velacreations.com.

DIY OUTDOORS

Wood Gas Camp Stove

Make a simple tin-can stove that costs 99 cents, runs for free, and sequesters carbon as you cook.

By William Abernathy

NEARLY HALF THE WORLD'S POPULATION cooks and heats using solid fuel, much of it burning up in pits that have seen no improvement since *Homo erectus* first tamed fire. This is not a small problem: inefficient cooking fires waste fuel, impoverishing both the planet and the person burning it; they inject startling quantities of soot, carbon dioxide, and worse greenhouse gases into the atmosphere; and they injure and kill the families who use them to cook and stay warm.

You can build a simple example of an appropriate technology that addresses all these problems: a biomass gasifier stove. It sounds more complicated than it is. Charring wood or other natural solid fuels releases gases that burn quite nicely. If you've ever

watched a campfire closely, you've seen little jets of smoke erupt from the wood ahead of the flame. If conditions are perfect, a smoke jet catches fire and turns, briefly, into a tiny geyser of flame. A good gasification stove recreates these conditions reliably, generating smoke and moving air to produce these little fire geysers on demand.

There are many designs for efficient stoves, and gasification is only one way to boost the efficiency of a cooking fire. The wood gas stove in this article is an elegantly simple gasifier design called a *TLUD stove* (for *top-lit updraft*), also known as an *inverted downdraft stove*. If you don't care how it looks, you can build it with a can opener, a punch, and a big rock. This design, which I've adapted from one

William Abernathy

I first saw on Instructables, is built around a 1-quart paint can. It easily boils enough water for a small pot of tea or a bowl of noodles, using nothing more than a fist-sized charge of scrap wood.

How It Works

This TLUD stove works in batches: fill it up with twigs and woody bits, and light it from the top. As the fire burns, it makes a layer of hot coals, and as this burning mass descends into the can, it becomes starved for air. Enough oxygen comes up from below to keep the embers alight, but not enough to sustain a flame.

This charring layer slowly descends, releasing flammable gases as it goes — a process called *pyrolysis*. The hot gases rise to the top of the stove, where they are met by an inrush of preheated air and, if all goes well, combine completely with this air in a clean secondary burn that consumes the methane, soot, and carbon monoxide produced by the primary combustion.

When a TLUD is dialed in, it's lovely: a layer of smoke hangs over the coal bed like a fog, and appears imprisoned by a gauntlet of inward-facing flame jets that rush in from the edges of the stove. While the stove is gasifying, it's remarkably clean: there's little or no smoke and only a faint odor, reminiscent of diesel or creosote. It leaves a trace of soot on your pot, compared to the heavy blackening an open fire imparts. And when your stove is done gasifying and the flame jets go out, clean charcoal remains in the stove.

This residual charcoal is a fringe benefit: not only can you use it for gunpowder (*see MAKE Volume 13, page 54*), but you can cook with it a second time in a clean-burning charcoal fire. You can also throw it into your compost and bury it. Called *biochar*, this buried charcoal enriches the soil and actually makes your carbon-neutral biomass fuel carbon-*negative*.

Make Your Gasifier Camp Stove

1. Prepare the backup block.

Open the paint can and drop the block of wood upright in the can (Figure A). Mark the

MATERIALS

Paint can, 1qt, steel, clean Hardware stores sell quart paint cans for about a buck. Don't use one that has held paint.

Food can, 19oz, steel Commonly used for baked beans, chili, and pineapple chunks, these cans snap neatly into the top of the quart paint can. The critical dimension, the lip of the can, must be very close to 3⅜". Do not remove the bottom.

Chicken or tuna can, 12oz, steel to make a standoff that keeps the pot from smothering the flame

Scrap wood, 1½"×1½"×7" (or longer) for a backup block, to keep the cans from denting under the drill, or spinning and hurting you if the drill bit grabs.

TOOLS

Saw (table or hand), or router	**Prick punch**
Computer, printer, and drilling templates download from makeprojects.com/v/27	**Hammer**
	Electric drill
	Step bit, ½" It's safer and cuts a cleaner hole than twist bits, which grab and tear sheet metal.
Caliper, ruler, or tape measure if you're not using the drilling templates. Electronic, dial, or Vernier calipers will work.	**Can opener**
	Tinsnips
	C-clamp
	Half-round file
Machinist's dividers with layout dye or a Sharpie marker; and woodworker's marking gauge or machinist's surface gauge if you're not using the templates	**Workbench**
	Vise
	Eye protection

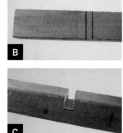

top edge of the can on the block.

Withdraw the block and make a parallel mark ⅜" below the first (Figure B).

Cut a ⅝" deep slot between these lines (Figure C). Precision is not critical: anything wider than ⅜" will do. A table saw is easiest, but a router or a handsaw and chisel will work too.

2. Mark and drill the 1qt paint can.

Use marking dye or a Sharpie to lay down a stripe ¾" from the bottom of the paint can (Figure D). Then, using either a surface gauge or a marking gauge, circumscribe a line ¾" from the bottom of the can (Figure E).

Clamp the backup block to the edge of your workbench with its slot facing up. Slip the paint can over the end of the block and drop its rim into the slot (Figure F). Find the can seam, and punch a point on the scribed line about ½" away from the seam (Figure G). This is your 12 o'clock mark.

If you have a caliper, set your dividers to 1.075" (¹⁄₁₂th division of the can in chord sections); if you don't, set your dividers to a hair more than 1¹⁄₁₆". Following the scribed line, use the dividers to scribe marks to the right and left of the initial punch mark, marking as you go, to the far side (Figure H).

When you reach the far side, your scribe marks may meet perfectly at 6 o'clock, but will likely over- or under-shoot each other. If you've "nailed it," punch it down; otherwise, split the difference and punch between the two marks.

If the cumulative error is not excessive, punch the other marks back to the first punch mark (Figure I); otherwise, adjust your dividers, re-scribe the ¾" line, and use the same method from 6 and 12 o'clock to mark and punch the 9 and 3 o'clock positions. When you like what you see, punch down the 8 remaining marks. Perfect spacing is not critical.

To speed things up, you can approximate this method by wrapping a tape measure around the circumference, or you can avoid measuring entirely by downloading the drilling templates from makeprojects.com/v/27 and gluing them to the cans (Figure J).

With the can resting on the backup, drill ½" holes at the punch marks. On your first hole, push only the tip of your step drill through the can and into the wood.

Remove the can from the backup block and put the drill tip back into the hole you just made. Drill straight into the wood past the ½" mark, creating a step-bit-shaped recess

(Figure K). (If you skip this step and drill straight through the can into the wood, your first hole will be egg-shaped.)

Next, steady the can with your hand, and allow it to "float" a little as the drill heads down its hole (Figure L). Don't force the drill: the step bit will cut effortlessly and cleanly.

When you're done drilling, wash off your marking dye or any glue. (Steel marking dye comes off with brake cleaner, Sharpie with rubbing alcohol, and most glues with lighter fluid.) When this can gets hot, any remaining goo will be cooked on forever.

Using the can opener, remove the bottom of the paint can.

3. Mark and drill the 19oz food can.

Using the methods described in Step 2, apply your drilling template, or mark and scribe a line ½" from the open end of the 19oz can (Figure M). Set your dividers to the chord division of 0.431" (a little less than ⁷⁄₁₆"), and make 24 divisions at the ½" line.

Punch your marks, and drill twenty-four ¼" holes (Figure N). Remember to create a ¼" step-drill recess in the wood before you drill all the holes.

Clamp the wood block vertically (a vise comes in handy here) and slide the 19oz can over the end (Figure O).

Drill at least thirty ¼" holes in the bottom of the can (Figure P). You can do a fair job by eyeball if you follow the rings stamped into the bottom of the can. The important thing is that you create an open enough bottom for good airflow while not letting all your fuel fall out. Clean off any dye or glue remaining.

Snap the 19oz can open-side first into the paint can. They should make a satisfying press fit (Figures Q and R).

4. Mark and drill the standoff.

Apply the template, or mark and scribe a line ¾" from the top of the 12oz can, and set your dividers to the chord division of 0.767" (a hair past ⁴⁹⁄₆₄"). Scribe 16 divisions at the ¾" line (Figure S). Start anywhere: these cans are seamless.

Punch your marks and drill sixteen ½" holes. Clean off any dye or glue.

Turn the can upside down, and drill a ½" hole through the bottom of the can (Figure T). Dig in one jaw of your tinsnips and cut around the hole, spiraling out to make a hole about 2½" in diameter (Figure U, following page). Use the ridges and valleys pressed into the bottom of the can as landmarks. When you're satisfied, turn the wood block up to vertical and hammer down any rough edges (Figure V), finishing the job with a half-round file.

Using the Wood Gas Stove

Fill your stove with wood chips, acorns, eucalyptus pods, small pinecones, or dried pelletized dung (Figure W). Your fuel must allow some airflow through it, so don't use loose sawdust or a single big chunk.

Starting these stoves is not easy — you may have to cheat with a little charcoal lighter, Sterno, or other fire-starting material (my favorite is old denim strips soaked in wax).

As the fire burns down past the side jets, blow down on the coals to start pyrolysis. Try to get a uniform coal bed all the way around the can. Once you see good coals with fire floating on top (Figure X), put the standoff on the stove, rim up (Figure Y). If you see an orderly column of flame rising with no smoke, you're gasifying! Put a pot of water on top and get your ramen on (Figure Z).

When the fire goes out, you're done with this charge. You can pile another charge on top, but you may need to relight it. To save the charcoal for reuse or sequestration, pour in some water. To reduce it to ash, leave it for another half hour and it'll burn away.

When you're done, the standoff nests neatly between the other 2 cans for easy storage.

You can experiment with changing the intensity and duration of the flame by changing the size of the hole on the top of the stove. The standoff lets you choose between a 2½" hole or a wide-open top simply by flipping it over. You can add old can lids with different-sized holes cut in the middle to "throttle" the flame higher or lower. ◪

U

V

⚠ WARNINGS! This stove produces and consumes carbon monoxide, a deadly poison. **DO NOT EXPERIMENT WITH THIS STOVE INDOORS!** Use it only outdoors.

Your stove will get hot. Using it on a wooden bench will leave a scorch mark. Use it on concrete, a tile you don't care for (thermal shock may crack it), or on dirt you don't mind scorching.

W

X

◪ TIP: Your gasifier will burn most of the soot, but not all of it! For easier cleanup, coat your pot with a thin layer of dishwashing liquid before you start. The soot will rinse off without scrubbing, and there'll be a small patch of burnt soap that comes off with a light scrub.

Y

Z

William Abernathy (yourwritereditor.com) writes and makes things in Berkeley under the watchful eyes of a feisty book-binder, two little girls, and one old cat.

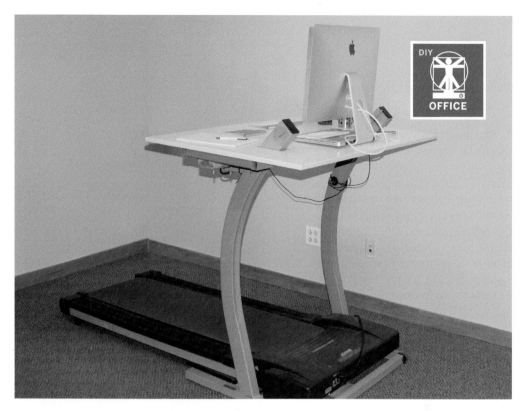

Treadmill Desk

An "active workstation" from a treadmill and Ikea parts.

By Doug Bradbury

A FEW YEARS AGO, WHEN MY SOFTWARE company was buying its first office furniture, I coveted a $4,500 adjustable-height desk with a treadmill underneath. I thought this would be a good way to change a sedentary job into an active one.

Later, when I saw desk designer Dr. James Levine's research on non-exercise activity thermogenesis (NEAT), I learned that he'd originally mashed together a treadmill and a laptop desk to test the idea, and he raved about the results. I decided to make my own.

1. Source the parts.

My first step was to acquire a treadmill. I didn't want to pay retail price for something I was going to tear apart, so I looked to Craigslist (craigslist.com), which was full of people unloading their treadmills. I found a

MATERIALS

Treadmill must have sturdy upright supports and support the iFit Chirp protocol. I used a Sears ProForm 830QT.
Tabletop Ikea Galant
Pull-out keyboard shelf Ikea Summera
Lumber, 2×4, 4' long
Bolts, ⅜"×3", with matching washers and nuts (4)
Cable, stereo, with ⅛" male plug

TOOLS

Screwdrivers
Drill and drill bit
Jigsaw
Ruler
Rotary tool with cutting wheel
Tape masking or packing

ProForm 830QT (a Sears brand) for only $140. For treadmill specs and reviews, the Treadmill Sensei website (treadmillsensei.com) was a great resource.

My next stop was the discount "as-is" section of Ikea, where I snatched a display-model Galant table, frame, and T-leg combo for $60.

2. Tear everything apart.

I began construction in the way that any good mash-up project begins: by tearing everything apart. I unscrewed and unplugged everything I could find on the treadmill until I was left with just the tread, the upright frame, and the touch-panel control unit (Figure A).

One main control cable ran through a leg of the treadmill from the control panel to the motor below. There was also an audio cable, a cable running to the hand-grip heart rate monitor, and an extra heart rate cable that led to nothing. The idea behind treadmill desks is to walk slowly for long periods of time, rather than to get your heart pounding, so I yanked out all the heart rate cables and sensors.

3. Mount the tabletop.

Prying the rubber handgrips off each side revealed horizontal supports at a good height for the table. So I drilled a hole through the posts and bolted on a 2×4 that was long enough to span the Galant table frame.

After removing the frame from the tabletop, I chose 2 holes on the frame that lined up well with the 2×4, drilled holes through the 2×4, and bolted the frame on top (Figures B and C).

Things looked good except for the 2 upright posts that would prevent the tabletop from being installed. I had 2 options: either cut off the posts or notch out the tabletop. My tools and expertise were better matched to notching the table, so I proceeded to carefully measure and cut the holes with a drill and jigsaw. Then I screwed the tabletop to the frame from below.

4. Reconnect the control panel.

It was time to re-establish control of the treadmill. First, I used a rotary tool to hack away at the plastic enclosure containing the touch panel and circuit boards until I had a box just large enough to house them.

A

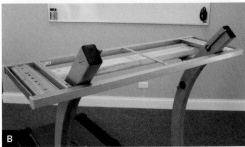
B

C

◥ TIP: When cutting through a laminated surface (like most Ikea stuff), stick masking or packing tape over your cut line to prevent the laminate from chipping.

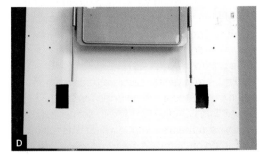
D

Conveniently, the box had 2 screw holes on the bottom, which I used to mount the box onto a $10 Ikea pull-out keyboard shelf (Summera), which I installed underneath the desktop (Figure D).

This way, I could easily pull the controls out when I needed to start and stop the treadmill or change the speed, then stow them away, to not interfere with walking or typing.

With the touch-panel enclosure mounted on the keyboard tray with its 2 original mounting screws, there wasn't quite enough slack in the control cable to pull the control display out far enough to see. So I took the cover off the treadmill base and found a way to reroute the control cable inside to free up sufficient slack (Figure E). After replacing the cover, the treadmill desk was complete.

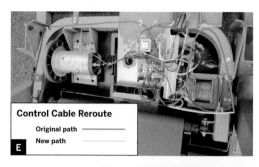

Control Cable Reroute
Original path ————
New path ————
E

5. Walk and code.

After building the treadmill desk at home, I packed it up and reassembled it in our office, where it continues to draw a lot of attention.

For a total cost of $210, it's been a great, healthy addition to our office and has succeeded in making my day a more active one. I comfortably type, code, Skype, talk, and do everything else that I did while sitting down. Now I just do it at 1 mph (Figure F).

One problem arose. We often write software in pairs, which helps us work faster and make fewer mistakes. But the treadmill desk only lets one person walk at a time. So we now have a double treadmill desk, custom-built from my design by a manufacturer in Indiana.

Software Upgrade

Being a software craftsman by trade, I soon became dissatisfied with reaching under the desk all the time to change the speed and incline of the treadmill. I had a computer on top of the desk, so why couldn't I control the treadmill from there?

Fortunately, my treadmill supports the iFit Chirp protocol, which is primarily used to enable a CD player with a workout CD to plug into the treadmill and send control commands in between music tracks.

F

The documentation on this protocol is sparse, but I found an open source project in C that generated the correct signals. I ported this library to Java and made it real-time. Then using my company's Limelight GUI framework for Ruby, my apprentice built an attractive user interface to drive the library.

After plugging the computer's audio output into the treadmill control input, I now had control of the treadmill right on my computer desktop. Instructions for downloading and running this app are at walkncode.com. ◪

Doug Bradbury is a software craftsman at 8th Light, Inc. who can't stop at just making great software. You can contact him via email (make@dougbradbury.com) or Twitter (@dougbradbury). He lives in the Chicago suburbs with his wife, Jen.

Touchdesk

Customize your own integrated workspace.

By Pauric O'Callaghan

AS A USER-INTERFACE DESIGNER, MY JOB is to understand people's goals and create a design that allows them to focus less on their tools and more on their work. I decided to apply this thinking directly to my own workspace. First I sketched out my basic design idea and defined my requirements, then I prototyped and iterated my design in electronics, software, and wood. The result is the Touchdesk, which I now use every day.

The keyboard and mouse are generalist input devices that enable all commands through the same interface, but the "80/20 rule" observes that we use 20% of features in any application 80% of the time. In my work, common actions require navigating to submenus or remembering multikey, app-specific shortcuts, while seldom-used commands use up dedicated one-button access.

To remedy this, I sketched out a smarter "soft" keyboard that runs on a touchscreen display inlaid into the desk's wooden top. Icons on the touchscreen offer one-button access to the actions I use most frequently, with the button mappings changing based on the current application. The buttons themselves graphically represent their functions. To zoom in, for example, you touch a magnifying glass icon instead of having to chord something like Ctrl + as you would on a keyboard.

Above the touchscreen, 4 inlaid wooden buttons are hardwired to Copy, Paste, Delete, and Ctrl, 3 commands and a modifier that are common across all applications and which I felt didn't need to take up virtual keyboard space. To the right of the touchscreen I inlaid a tablet, the preferred pointing device (replacing a mouse) for drawing and design work.

MATERIALS

Resistive touchscreen, 4-wire, 6½" or 7" such as part #360-2446-ND ($38) or #BER277-ND ($59) from Digi-Key (digikey.com). I purchased mine on eBay for around $20.

USB gamepad, one-hand keyboard with keyboard mapping software I used the 15-button Belkin Nostromo n52, $75; the current version is the n52te.

Pointing device for computer: tablet or mouse I used a Wacom tablet, but use whatever you like best.

Digital picture frame, 7"

Arduino Mega microcontroller (or Illuminato) part #MKSP5 from Maker Shed (makershed.com), $65. Standard Arduinos don't have enough inputs; I used an Illuminato, but the newer Mega will also work.

NPN transistors (15) any basic NPN, such as a 2N3904, part #COM-00521 from SparkFun Electronics (sparkfun.com)

Pushbutton switches, mini SPST (6) such as SparkFun #COM-00097

Protoboard or stripboard to make a shield for the Arduino, such as the 3"×4" 1200D epoxy fiber board from Veroboard (veroboard.com). I used a commodity pad-per-hole protoboard, 24×30 holes, 70mm×90mm.

Protoboard or stripboard for mounting 4 of the mini switches, such as Veroboard's 2"×10" 2000L boards (epoxy or phenolic). I cut a piece (around 8×60 holes) from some stripboard I got from an old lab.

USB hub, 4-port

Wire, 22 gauge insulated, various colors

Wood glue

Wood, 1"×12", 4' long I used pine for prototyping, oak for the final build.

Hardwood slat, ⅛" thick, at least 1½"×2' such as #64823 from Rockler (rockler.com)

Wood screws, 1¼" (4) #8 screws or whatever you have on hand that will work

Wood screws, ½", self-tapping (6) #6 or whatever size will work

Clear polyurethane spray, semigloss

Male breakaway headers (56 pins total) SparkFun #PRT-00117 (2x)

TOOLS

Plunge router with round-over and mortise bits
Drill and drill bits (for wood screws)
Screwdriver
Glue gun and hot glue
Ruler and pencil

Tools I used for prototyping (which you won't need if you copy my design):
Boarduino Kit Maker Shed #MKAD9, $18
Arduino Mini USB adapter Maker Shed #MKSP3, $20
Plug-in breadboard power supply, 3.3V/5V Maker Shed #MKSF5, $14
Solderless breadboard
USB-TTL serial cable, 5V

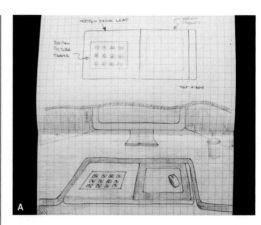

A

I also wanted to make sure the physical desktop had room to accommodate a paper notepad, along with a traditional keyboard (which I inset in wood to match the desk) that could stow away and be brought out for occasional text entry and document writing.

At the start of the project I wrote out a list of all my design requirements and sketched out the physical workspace (Figure A). I recommend this practice for any major project as a way to stay focused on your vision and decrease the chance that you'll lose interest and shelve it for another day.

System Architecture

Figure B (following page) represents the Touchdesk's functional architecture, showing how its input and output devices all connect to the computer via a shared USB hub. The touchscreen display has 2 parts: a standard digital picture frame underneath that shows the icons, and a transparent resistive touchscreen on top that reads the finger presses.

The Arduino drives the resistive touchscreen by applying 5 volts across it, alternating between horizontal and vertical. If a point is being touched, the screen will return a voltage between 0V and 5V for each axis, depending linearly on the point's distance from the screen's left (with horizontal voltage applied) or bottom (with vertical voltage applied). Pressing the exact center, for example, returns 2.5V for both x and y.

The x and y voltages from the touchscreen run to analog input pins on the Arduino. The

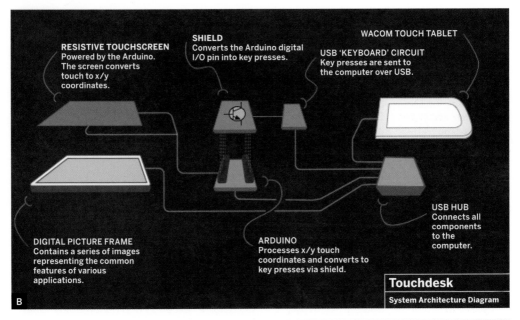

RESISTIVE TOUCHSCREEN
Powered by the Arduino.
The screen converts
touch to x/y
coordinates.

SHIELD
Converts the Arduino digital
I/O pin into key presses.

WACOM TOUCH TABLET

USB 'KEYBOARD' CIRCUIT
Key presses are sent to
the computer over USB.

USB HUB
Connects all
components
to the
computer.

DIGITAL PICTURE FRAME
Contains a series of images
representing the common
features of various
applications.

ARDUINO
Processes x/y touch
coordinates and converts to
key presses via shield.

Touchdesk

System Architecture Diagram

B

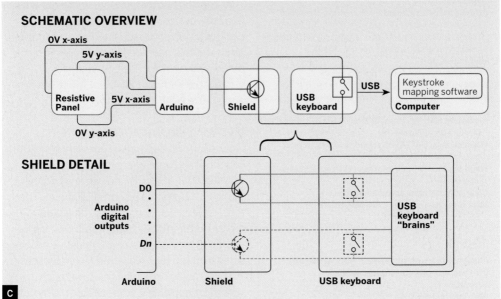

SCHEMATIC OVERVIEW

0V x-axis

5V y-axis

Resistive
Panel — 5V x-axis — Arduino — Shield — USB keyboard — USB — Keystroke mapping software / Computer

0V y-axis

SHIELD DETAIL

D0

Arduino
digital
outputs

Dn

Arduino   Shield   USB keyboard

USB
keyboard
"brains"

C

Arduino code then maps these coordinates to the button currently displayed at that position and turns on one of its digital output pins to match.

Each button on the touchscreen maps to one of the Arduino's output pins, which is why this project requires an Arduino Mega or Illuminato; standard Arduinos don't have enough I/O to support more than a small number of touchscreen keys, and the Touchdesk uses up to 15 inputs: 11 soft keys plus 4 hardwired buttons.

To convert the Arduino outputs into something the computer and my applications can interpret, I used a USB gamepad with 15 keyboard buttons. Pressing a button on the gamepad closes the connection between 2 points on its internal circuit board, so I wired the Arduino to trick the USB keyboard circuitry into thinking the user pressed a key.

Each Arduino output connects up to an NPN transistor on a plug-in Arduino shield PCB that I built, and each transistor connects out to the USB gamepad's PCB (Figure C). When a normally open transistor on the shield receives a digital HIGH from an Arduino pin, it closes the transistor gate, making the connection between a button's contact pair. To the computer on the other end of the gamepad's USB cable, this looks exactly like a key press.

Above the touchscreen, the hardwired Copy, Paste, Delete, and Ctrl buttons connect directly to contact pairs on the gamepad, in parallel with the Arduino-controlled transistors.

The final piece of the puzzle is configuring the gamepad driver software running on the computer to translate incoming key-press signals into the appropriate actions. The Belkin Nostromo n52 has a nice user interface for making these associations, and it can store a unique mapping for each application (Figure D). The driver then automatically switches mappings to match the app that's currently running. For example, when I press the top left button on the touchscreen while I'm using Photoshop, I get the brush tool, but if I move over to Word, that button summons the highlight tool.

There is one limitation: although the touchscreen's button mapping changes functionally when I switch applications, I didn't figure out an automatic way to tell the display which application is running. For switching the button icon arrays to match different applications, I wired 2 pushbutton switches to the picture frame's Next and Back buttons.

After I began using the Touchdesk, my fingers soon learned the soft button locations for each app, so I don't generally rely on the icons anymore anyway. But the ability to manually switch screen images lets me bring up the right set when I need to.

Touchscreen Electronics

Four-wire resistive touchscreens are simple: one wire runs to each side, and you use the

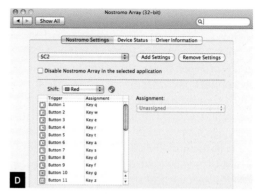

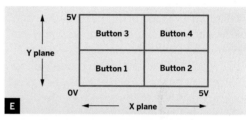

same wires to both apply voltage and take a reading along each axis. To hook an Arduino up to the screen, you connect one digital I/O pin to each screen wire for applying voltage, plus one analog input pin each to the screen's top and right-side wires.

I modified a sketch by Marco Nicolato to program the Arduino to read my touchscreen and determine which button is pressed;

you can download the code at makeprojects. com/v/27. For these experiments I used an Arduino Diecimila and Boarduino (Figures E and F, previous page), although I knew that this project needed a microcontroller that could support more dedicated outputs. With the digital picture frame, I simply removed its workings from its case.

I built the Arduino shield on a small proto-board, arranging male pin headers to plug into the Arduino and arraying the top with transistors. Then with the USB gamepad, I extracted its circuit board from its case and traced back contact pairs from each physical key location to find places I could solder to. After wiring each transistor on the shield to a contact pair on the gamepad PCB, I hot-glued the gamepad PCB to the shield (Figure G).

To connect the picture frame board, Arduino, USB gamepad board, and tablet to my computer, I used a 4-way USB hub.

Woodwork

"To carve an elephant from a block of marble, remove everything that doesn't look like an elephant." This old saw sums up how I made the Touchdesk's case. I took some wood and cut out sections to fit the internal electronics and keyboard on top.

I built a prototype using soft pine to figure out dimensions and ergonomics (Figure H). This taught me that I needed to use a smaller tablet and align the touchscreen with my left hand (I'm right-handed) rather than my face.

For the final design, I glued and screwed together a 2-layer case out of 1"-thick oak board. I carved the bottom board with a plunge router to resemble a tray. I carved the top layer to match, creating a hollow cavity for the electronics.

Then I marked and cut a hole to fit the touchscreen components, which I hot-glued to the inside of the case. For the tablet, which is thin, I used the plunge router to carve an indentation just deep enough to let it sit flush with the desk's top surface. Figure I shows the final woodwork, and Figure J shows the case with electronic components placed.

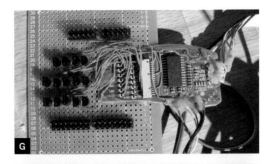

⚠ CAUTION: Routers are powerful tools that can easily run through the wrong part of the wood, your fingers, or both. Take care and time with them, and practice with a prototype before working on the final design.

Physical Buttons

Above the touchscreen I cut a hole for the 4 hardwired wooden buttons (Copy, Paste, Delete, and Ctrl). These buttons consist of thin, flexible, hardwood slats with pushbutton mini switches on stripboard (screwed to the bottom board underneath) that are wired via a pin header to the Arduino shield (Figure K).

By carrying 4 very common functions, these wooden buttons reduce wear and tear on the touchscreen and add a nice aesthetic to the overall design.

After wiring the 2 pushbutton switches to the digital picture frame's Next and Back button contacts (for changing the touchscreen icons), I hot-glued them under the desktop along the left edge, just behind the touchscreen (Figure L, upper right).

Final Design Recursion

Finally, with the software stable and the wood finished with clear polyurethane, I used the Touchdesk itself to complete its own design. On the Touchdesk, I created the touchscreen icon array images, which I saved onto an SD card and plugged into the digital picture frame.

Touchdesk 2

Having used the Touchdesk for more than a year now, I've compiled a list of improvements for the next version. One problem is that the heat from all the electronics inside the case dried and shrank the wood, causing the resistive screen to ripple and warp. I had to remove the picture frame and reglue in the resistive screen.

Again, the main technical limitation is my having to manually change the images on the digital picture frame to match the application on-screen. It's a small issue, but I would like to find a way to get that information from the computer to the display.

With that said, the exercise of building my own input device and integrated workspace gave me a better understanding of ergonomics and touchscreen technology and some good coding experience. ◪

J

K

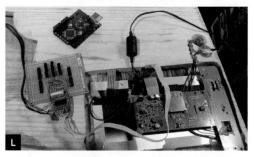

L

M

➕ Download the Arduino code file *touch_read.pde* at makeprojects.com/v/27.

Pauric O'Callaghan (pauric@pauric.net) is a user-interface designer by day and a carpenter/maker/hacker/kitesurfer by night.

The Do-Not-Touch Box
Use a simple accelerometer to perplex your friends.

ACCELEROMETERS HAVE BECOME ubiquitous. Your car contains an accelerometer that triggers the deployment of an airbag if it senses a collision. The game controller on a Nintendo Wii contains an accelerometer that responds to your physical movements. If you build a robot, and you want it to react appropriately when it bumps into something or falls over, an accelerometer can take care of that, too.

Modern accelerometers are truly amazing, microscopic devices, packaged inside tiny integrated circuit chips that can cost less than 50 cents each. This kind of chip is known as a MEMS (micro-electro-mechanical system), and if you want to see how they work, you can find some fascinating electron micrographs by searching online for "mems device."

I can imagine dozens of hobby projects using accelerometers. The only problem is that MEMS chips are so small, typically less than 5mm square, that they're very difficult to work with. Fortunately you can buy one already mounted with associated hardware on a mini-board measuring ½"×¾", ready to be plugged into an everyday breadboard.

The price may seem high (the one I recommend currently retails for $23, which is among the cheapest) but in the project I have in mind, the additional parts should cost you less than $5.

Experimenting with the Accelerometer
I selected the DE-ACCM6G made by Dimension Engineering (Figure A). This is a 2-axis accelerometer, meaning that it will detect

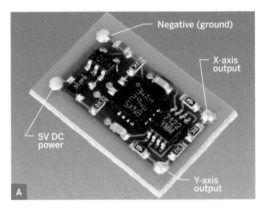

Negative (ground)

X-axis output

5V DC power

Y-axis output

A

transitions in 2 perpendicular directions. Looking down at it from directly above, when you move it from left to right, this is considered motion along its x-axis, and the voltage at the lower-right pin will increase momentarily. Move it from right to left, and the voltage on the same pin will diminish momentarily (see Figure B).

When you push the accelerometer away from you, and then pull it back toward you, this is considered motion along the y-axis, creating voltage fluctuations on the lower-left pin.

You can easily test this with just a 9V battery, an LM7805 voltage controller, and a multimeter, as shown in Figure C. The LM7805 supplies 5V, which you connect with the 2 top pins on the mini-board. But which end is the top end? You can compare it with the photographs here, or look underneath, and you should find the pins identified as X and Y (the outputs), Gnd (negative voltage), and Vin (positive voltage). Make sure you apply the voltage to the correct pair of pins, the right way around!

Charles Platt

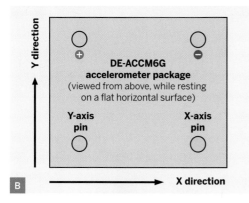

Y direction

DE-ACCM6G
accelerometer package
(viewed from above, while resting
on a flat horizontal surface)

⊕

⊖

Y-axis
pin

X-axis
pin

B

X direction

MATERIALS

DE-ACCM6G accelerometer available from
trossenrobotics.com or robotshop.com
Resistor, ¼W or higher, 3.3kΩ
**Piezoelectric beeper, steady-tone, drawing less
than 20mA** such as Kobitone 254-PB511-ROX or
model 273-073 from RadioShack (radioshack.com)
Differential comparator such as TLC372IP,
TLC372CP, or similar
LM7805 voltage regulator You can ignore any
additional letters appended to this part identifier.
9V battery and snap connector
Breadboard and jumper wires

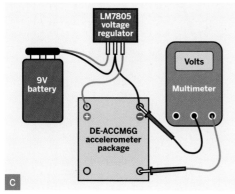

LM7805
voltage
regulator

9V
battery

Volts

Multimeter

⊕

DE-ACCM6G
accelerometer
package

⊖

C

✎ **Fig. A:** (Opposite) Closeup of the DE-ACCM6G
accelerometer package, mounted on a mini-board
measuring ½"×¾". Four pins (unseen, below the board)
insert into a standard breadboard. Solder pads above
each pin are labeled with the pin functions.

✎ **Fig. B:** Functions of the output pins of the accelerometer when it's moved in the directions shown.

✎ **Fig. C:** Necessary connections to test the accelerometer. In reality it should be mounted with the voltage
regulator on a breadboard, using jumper wires to
make the connections shown.

Set your meter to DC volts, using the 2V
range if the meter doesn't do auto-ranging.
Hold the black probe in contact with the negative side of your power supply, and touch the
red probe to each of the little solder pads above
the output pins of the accelerometer. While
the component is horizontal and motionless,
your meter should show approximately 1.65V
on each pin. Now slide the accelerometer
around, and watch how the voltages change.

What happens if you lift the mini-board
directly upward? Nothing! To detect motion in
the third dimension, you would need a 3-axis
accelerometer.

Try tilting the breadboard away from you
until it's vertical, resting on its top edge.
After the Y output stabilizes, it should show
about 1.88V, suggesting that the component
is accelerating downward, even though it
isn't moving. How can this be?

Well, one of Einstein's great insights was
that gravity is indistinguishable from acceleration. You could be sitting in a rocket out
in space, accelerating smoothly at 1 gravity
(1g), or you could be sitting motionless in that
same rocket on the launch pad, experiencing
the gravitational pull of the Earth, and you
would have no way to detect the difference.
Strange but true: you and all the objects
around you are experiencing a constant 1g
downward acceleration right now.

Here's another little experiment. Hold the
accelerometer, the meter, and the battery on
your knees while you spin around in an office
chair. The accelerometer will measure your
outward centripetal acceleration, which is
often referred to (inaccurately) as centrifugal
force. Acceleration is a fascinating concept,
and your accelerometer can help you to
explore it.

But how can we make practical use of the
chip on its mini-board? The smart way is to
connect it with a microcontroller such as the
Arduino, BASIC Stamp, or Picaxe. All of these
chips contain analog-to-decimal converters,
allowing you to write a little program that
tells the microcontroller what to do when it
senses small changes in voltage. You can set
up a numeric display showing acceleration
in gravities (g), and mount it in a car, on a
bicycle, or on your own body when you jump

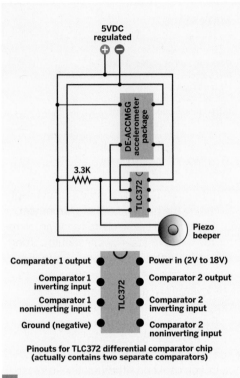

5VDC
regulated

DE-ACCM6G accelerometer package

3.3K

TLC372

Piezo beeper

Comparator 1 output ●	● Power in (2V to 18V)
Comparator 1 inverting input ●	● Comparator 2 output
Comparator 1 noninverting input ●	● Comparator 2 inverting input
Ground (negative) ●	● Comparator 2 noninverting input

TLC372

Pinouts for TLC372 differential comparator chip (actually contains two separate comparators)

D

E

up and down on a trampoline.

But let's start with something simpler. I'm going to use the accelerometer to make a Do Not Touch Box.

→ START

Building the Box

I have to admit that if I see a little box with "do not touch" written on it, I can't resist picking it up. This little project will deliver a harmless surprise to anyone who yields to that temptation.

In Figure D, the schematic shows that 2 outputs from the accelerometer can drive the 2 inputs of a chip called a comparator (part #TLC372). One of these inputs is known as the inverting input (usually marked with a minus sign) while the other is called the non-inverting input (marked with a plus sign). The comparator constantly compares its inputs. If the inverting input has a higher voltage than the non-inverting input, the output from the comparator will be low, and vice versa.

✎ **Fig. D:** Schematic showing a TLC372 comparator attached to the accelerometer package, driving a beeper. Because the TLC372 output goes low when active, the beeper is connected to positive power, which it sinks into the comparator. The 3.3K resistor pulls the output of the TLC372 higher when it's not sinking current.

✎ **Fig. E:** This breadboard layout corresponds with the schematic in Figure D. Wires from a 9V battery are applied to the top 2 rows of the breadboard. The LM7805 voltage regulator, seen from above, is the black rectangle at top right. The TLC372 comparator pinouts are: pin 1, output; pin 2, inverting input; pin 3, non-inverting input; pin 4, negative power; pin 8, positive power; pins 5, 6, and 7 connect with a second, separate comparator inside the chip, and can be left with no connection.

Because the comparator has a very high input impedance, you can wire it directly to the accelerometer mini-board.

The comparator is an open-collector device, which means that you should use its output pin to sink current. (To learn more about this concept, consult an introductory book, such as my own *Make: Electronics*.) Therefore, you can think of the comparator as "switching on" when its output goes low. It can sink 20mA, which is easily enough to

power a piezoelectric beeper.

Connect positive voltage to the red wire of the beeper, connect its black wire to the output from the comparator, and the beeper will beep when the comparator output goes low. However, you also need to use a pull-up resistor to insure that the comparator output is high when the comparator is inactive (Figure E).

The beeper may not beep when you first apply power to the circuit. Gently tilt your breadboard in each direction, and whenever the x-axis output is greater than the y-axis output, the comparator will activate the beeper.

Inside a suitable box, mount the components at a slight angle, so that the beeper will remain silent while the box is resting on a level surface. Label the box, leave it in an obvious place, and wait for someone to take the bait. As soon as the box is disturbed, it will start beeping, and the person holding it will have a hard time figuring out what makes it beep and what keeps it quiet.

Because the electronics consume only about 5mA while they're waiting for something to happen, the 9V battery should last for 2 or 3 days of continuous use.

Going Further

If this little project rouses your curiosity, here are some more ideas:

» Search online for *comparator hysteresis* to learn how a comparator can tolerate larger differences between inputs before emitting a signal. Hysteresis is an important concept when sensing changes in the environment.

» Connect the x-axis output from the accelerometer to one input of the comparator. Using a voltage divider, apply a fixed, constant 1.65V to the other input. Use hysteresis so that the comparator isn't triggered unless the voltages on its inputs differ significantly. Now the comparator only senses acceleration or tilting along the x-axis.

» The TLC372 chip actually contains 2 comparators. You can use the second one to sense changes in the y-axis. Now you have the beginnings of an attitude control system in some kind of flying or walking device.

» You can wire the comparator so that it only responds to large fluctuations. In this mode, it will detect impacts or collisions. You'll find a circuit that does this on the Trossen Robotics website, which sells the DE-ACCM6G comparator (see the Materials list).

» Try using a 555 timer to generate an audible tone, and feed an output from the accelerometer to pin 5 (the Control pin) of the timer. When you hold the accelerometer while making hand motions, the tone will change pitch. (You may need an op-amp to increase the accelerometer voltage output for the 555 input.)

I'm sure you can come up with many more fun accelerometer projects of your own. ◪

➕ *Make: Electronics* book at the Maker Shed: makezine.com/go/makeelectronics

Charles Platt is the author of *Make: Electronics*, an introductory guide for all ages. A contributing editor of MAKE, he designs and builds medical equipment prototypes in Arizona.

Hot Water!

The Scenario: You've just returned to your country home from a shopping trip to town, some 20 miles upriver, with all the fixings for a weekend barbecue with your family, including a fresh bag of your favorite mesquite charcoal and some other sundries: coffee filters, kitty litter for your pet Persian, Sheba, and some fresh aquarium sand for your daughter's collection of goldfish — all of whom she's named Moby Dick. No sooner are you in the door than the power goes out. And, calling the electric company to report the outage, you quickly discover the loss of power is now the least of your worries.

There's been a serious accident at the nuclear power plant 5 miles upriver from town, and radioactive steam is now escaping into both the air and the river. Containment of the leak and restoration of power are both currently indefinite. You are advised to stay indoors and drink only bottled water until further notice, as they now presume the river water is contaminated well past your location.

The Challenge: You consider getting everyone into the car and evacuating. But, given the prevailing winds, your only way out would certainly put you right in the path of the leak — so that's really not an option. And while your large house can be sealed up easily — and there's enough food for a week or more if need be — aside from some beer and a few bottles of soda, you have no bottled water! So what are you going to drink? You must devise a way to provide enough safe drinking water for your family of four to weather the crisis for at least a week.

What You've Got: In addition to everything mentioned, you have a garage full of tools, a 5-gallon plastic jug your family uses to collect spare change, and anything else that would normally be found in a typical house. So prepare to hunker down, break out the board games, and protect your nuclear family. Good luck (to all of us).

Send a detailed description of your MakeShift solution with sketches and/or photos to makeshift@makezine.com by Oct. 28, 2011. If duplicate solutions are submitted, the winner will be determined by the quality of the explanation and presentation. The most plausible and most creative solutions will each win a MAKE T-shirt and a *MAKE Pocket Ref*. Think positive and include your shirt size and contact information with your solution. Good luck! For readers' solutions to previous MakeShift challenges, visit makezine.com/makeshift.

Lee David Zlotoff is a writer/producer/director among whose numerous credits is creator of *MacGyver*. He is also president of Custom Image Concepts (customimageconcepts.com).

REMOVE THE *HORNS*

INSERT THE SERVOS *BACK TO BACK* SO THE SHAFTS ARE NEAR THE BOTTOM

MARK THE TUPPERWARE TO INDICATE *WHERE* TO MOUNT THE SHAFT

DRILL THE HOLES

MOUNT MOTORS USING *ZIP TIES*

CONNECT THE WIRES AS SHOWN *BELOW*

PLACE THE BATTERY PACK *INSIDE* AND REPLACE THE *HORNS*

CUT OFF COAT HOOKS

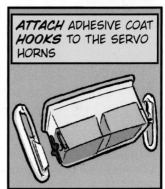

ATTACH ADHESIVE COAT *HOOKS* TO THE SERVO HORNS

POWER IT UP!

WE *DID* IT. WOBBLE! WOBBLE!

BURP!

WE MAKE A GREAT *TEAM!*

Multimeters and soldering irons and scissors, oh my! Plus a reference manual for pyros, and the ultimate maker belt.

TOOLBOX

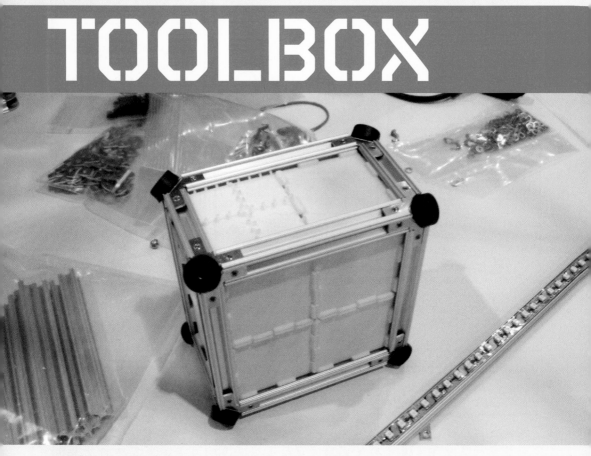

MicroRAX
Starter Kit $80; Pro Kit $180 microrax.com

Need miniature aluminum girders for that next robot project? Look to Twintec, a tiny company based in Washington state that manufactures and sells the lightweight MicroRAX building system.

Parts include 10mm extruded aluminum beams and all the braces, plates, brackets, and other connectors you need to make your masterpiece. They've even developed adapter plates to attach your 'RAX to VEX and NXT constructs, allowing you to merge multiple building media.

The beams, Twintec's signature component, are sold in 900mm lengths you can cut to size based on your project (precut lengths are also available). This seems eminently sensible —

forget futilely pawing through a box of wrong-sized beams for that right one, and hack your own.

At Maker Faire the company was showing off some awesome new additions to the line, including a swiveling plate (kind of like a hinge) they used to build a robotic hand. They also showed how to make 3D-printed panels that slide into the girders' grooves to form an enclosure. It's a young product, so there's still a lot you can't do with MicroRAX, but you'll be surprised at what you can build: anything from mundane shelving to robots to computer enclosures. For those super-elite projects, you can buy MicroRAX in ninja-black anodized aluminum!　　　　　　　—*John Baichtal*

Desktop Trebuchet Model Kit
$40 rlt.com/10421

RLT Industries has gone above and beyond with this kit. Each piece of wood is precisely laser cut. All the required tools can be found around your home or at a hobby shop, and the instructions are well-documented and detailed. However, as instructed, I used wood glue but it didn't hold well. So I resorted to a hot glue gun.

The overall building process took about four hours. The Desktop Trebuchet kit comes with four wooden balls to fire, and it shoots about 15 feet. I showed it to my engineering club, and we fired circus peanuts with it. (One of my club members ate all the peanuts.) It's a great kit for the model hobbyist who's just starting out.

—Robert M. Zigmund

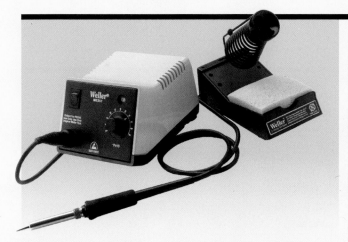

Weller WES51 Soldering Iron
$95 amazon.com

The phrase "you get what you pay for" must have been coined by someone weaned on a $9 Shack soldering iron, who scorched through a dozen electronic kits before realizing there's a better way: this gorgeous Weller.

The WES51 is a professional's tool, made to be used all day. It powers up quickly, with a range of 350°F–850°F. If you leave it on, the Weller automatically powers down after 99 minutes of inactivity.

Its comfy handle is made to be easy on the hands, and the ESD-resistant material keeps the shocks away from your project. Advanced users will appreciate its wireless temperature lockout, which limits the operator's ability to accidentally use too much heat. I don't know if it's the sum of these features that makes the Weller superior to the el-cheapo model, but I know it's true that I've never soldered better.

—JB

Toribe Kitchen Scissors
$42 niwaki.com

You would imagine that scissors were a done deal — nothing left to improve. That is, until you try these stainless steel kitchen scissors from Niwaki. The shear (ha-ha) genius of the Toribe scissors is their simplicity, an elegance only Japanese tool fetishists seem able to achieve.

The design allows for one wonderful feature: separation of the two blades so they can be washed (who wants last night's chicken guts sitting on his scissors?). This feature also enables easier sharpening of the blades. These are scissors for life.

—Saul Griffith

GorillaTorch Flare

$35 joby.com

The flexible, hands-free GorillaTorch Flare from Joby combines a 100-lumen LED flashlight with the Gorilla line's signature magnetic tripod. True to the name, the Flare is equipped with both white and red LEDs, combining a work light with a powerful emergency signal. Less dramatically, I've used mine as a miniature photo light, perched over a project so I can snap some well-illuminated shots.   —JB

Kindle
E-Reader

$140 amazon.com

The Kindle is a joy. The E-Ink screen has high readability in sunlight, and we makers can hack it to give our laptops and iPods internet access — although this isn't necessarily condoned by Amazon.

It's fairly drop-proof, but if you do drop it, call Amazon Kindle Support, and they'll say, "No problem. We'll overnight you a new one." As an added bonus, the MAKE PDFs show up great on mine, but since it's monochrome the pictures lose some of their bang.

—Zach Zundel

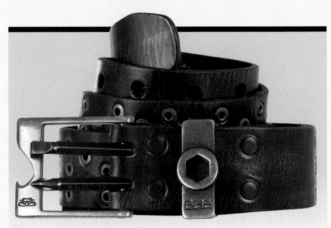

Tool + Belt

$40 686.com/store

This belt is the first of its kind, a utilitarian-yet-stylish maker-wear belt. Rather than the unwieldy, too-large carpenter's belt, the 686 belt, which I've affectionately dubbed "Toolbelt 2.0," is almost indiscernible from a regular leather belt. But there are key differences: its buckle consists of two screwdrivers, flathead and Phillips, and a bottle opener cut into the side, while its metal belt loop has three wrenches cut into it.

The leather belt, which comes in black, brown, and white, is designed so you can easily and quickly remove both tools. It's durable, waterproof, and, quite frankly, looks pretty cool, all reasons why it's almost always on my body. All in all, it's by far my favorite article of clothing and my second favorite tool (next to my trusty Leatherman Wave, of course).

—Adam Zeloof

Eureka! By Roy Doty
Phffft!

TORCH STANDOFF

These two torches from BernzOmatic (bernzomatic.com) are actually quite different. The QuickFire torch is a fantastic tool for soldering copper plumbing fittings or heating heavy-duty nuts and bolts to free them up. The cutting/brazing torch, on the other hand, provides the higher temperatures required to silver solder, braze, and do thin-gauge cutting, all of which are outside the bounds of the QuickFire. The cutting/brazing torch requires more skill and practice to master, whereas the QuickFire is dead simple and a bit of a brute-force tool.

—*Chris Singleton*

BernzOmatic QuickFire Torch and Fuel Cylinder
$70

Although it hasn't replaced my hobby knife, sanding block, or swivel vise, this torch has joined them as a permanent member of the tools left out on my bench at all times.

With its shorter and more stable design, it's a vast improvement over the propane torch that's been around for years. Its nicest feature is instant lighting at the pull of a trigger, which eliminates the need for two-handed lighting with a sparker; this greatly simplifies intermittent heavy-duty soldering and heating.

Additionally, the torch has a new adjustable burner design to provide more efficient heat for the specific job at hand.

BernzOmatic OX2550KC Cutting, Welding, and Brazing Torch Kit
$100

This torch is a great intermediate tool positioned somewhere between a standard propane torch and a full-blown oxyacetylene rig, and while it won't braze or blast through heavy-gauge steel, it will cut through thin material, braze thin-gauge ferrous (steel) material, and weld aluminum. It also lets you do precision silver soldering more easily and better than with a standard propane torch.

Although it took a little practice to get the hang of adjusting it properly, this torch did an excellent job, and my silver solder joints have greatly improved.

Make: TIPS!

Disc Sander Cover
I don't often use the disc section of my big belt/disc sander, so I made a simple cover so I don't have to worry about things (or me) falling against the rotating disc when I'm concentrating on using the belt.

Lubricating Saw Blades
Before cutting fret slots, or much of anything else, rub an old candle along the edge of your saw blade. You'll be surprised how much easier it makes the process to have a bit of lubrication.

—*Frank Ford*

Find more tools-n-tips at makezine.com/tnt.

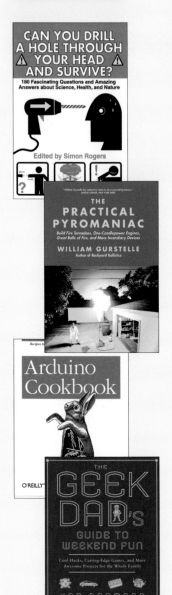

Drill Into Your Head

Can You Drill a Hole Through Your Head and Survive? by Simon Rogers
$13 Skyhorse Publishing

Here's a fun, interactive book that features 180 fascinating questions and answers about a wide range of topics, including Science and Technology, Sports and Games, and Health and Fitness. Each answer is backed up by a paragraph or two giving a thorough explanation using scientific research.

You'll be surprised by what you'll find out from this book, including my favorite questions: *Could we actually build a Star Trek phaser gun?, Is suspended animation safe?,* and *Are cloned animals safe to eat?*

Whether you're looking for a short, entertaining read that will open your eyes to some of the scientific research being done around the world, or just a book to put on your coffee table, *Can You Drill a Hole Through Your Head and Survive?* is a good choice.

—Kindy Connally-Stewart

The Wonders of Fire

The Practical Pyromaniac
by William Gurstelle
$17 Chicago Review Press

In his latest book, engineer and contributing MAKE editor Bill Gurstelle offers up 25 DIY projects that introduce the concepts and science of fire.

Gurstelle recounts theories about fire and tracks its evolution within the field of science. The book was written with the intent of using everyday household items and inexpensive materials; the first chapter includes a list of suppliers and a safety crash course in fire extinguisher use.

The projects range in level of difficulty so that as you progress through the book you accumulate pyrotechnic skills and move on to builds that progressively require a more elaborate setup. Several projects have been featured in MAKE, such as Flame Tube (Volume 26), Arc Light (Volume 20), and Fire Piston (Volume 19).

Other projects include the Burning Ring of Fire (a portable backpacking stove), the Fire Tornado, a DIY Thermocouple, and the large Propane Flame Thrower (seen on the book's cover). Full of historical detail, diagrams, and links to online videos, Gurstelle's book is the perfect reference manual for all things pyrotechnical. —Nick Raymond

Kitchen Coding

Arduino Cookbook by Michael Margolis
$45 O'Reilly Media

The Arduino development platform has taken the DIY community by storm. It's quite possible to go from zero knowledge of hardware and software dev to a blinking light in minutes. And while there are many online tutorials, they are often spread over different websites. The *Arduino Cookbook* creates an alternate learning path that's nicely packaged and complete within one book.

The *Cookbook* starts from the very beginning, detailing the Arduino hardware, integrated development environment (IDE), and C programming language, then jumps straight into writing code and making circuits.

The best part of this book (aside from its treasure trove of great code examples) is that each section starts with a problem followed by a solution and a discussion format.

Overall, this is a great book for beginners and intermediate microcontroller hobbyists and developers. The quick and clear instructions will get you hardware-hacking very quickly, and if you're like me and have worked with the Arduino for a few years, you'll still be pleasantly surprised with all the higher-end instruction. —Riley Porter

The Nerdy Family

The Geek Dad's Guide to Weekend Fun by Ken Denmead
$18 Gotham

Ken Denmead is *Wired's* resident GeekDad, and his second book of geeky projects is a wacky, fun, and inspirational collection. From making a Nerf dart blowgun or a trebuchet out of Legos to building robots from scratch or making stop-motion movies, there's a wide range of projects that will appeal to kids of all ages. The projects range from simple to moderately complex, and many involve recycling, reusing, and hacking things you'll already have around the house.

Most can be completed on a weekend afternoon, so if you're looking for some quick and educational activities, there's plenty in here to thrill your kids and the kid in you. Besides Denmead's inspired activities, there are also contributions here from GeekDad emeritus Chris Anderson and other guest authors like Rod Roddenberry, keeper of the *Star Trek* flame. —Bruce Stewart

New from MAKE and O'Reilly

MAKE: Arduino Bots and Gadgets
by Kimmo Karvinen and Tero Karvinen
$28–$38 O'Reilly Media

Want to build your own robots, turn your ideas into prototypes, control devices with a computer, and make your own cellphone applications? It's a snap with this book and the Arduino open source electronic prototyping platform.

Getting Started with the Internet of Things by Cuno Pfister
$20–$28 O'Reilly Media

The Internet of Things consists of billions of embedded computers, sensors, and actuators all connected online. This hands-on guide shows you how to start building your own fun, fascinating projects.

The Fun of Physics

Physics for Entertainment
by Yakov Perelman (translated from the Russian by Arthur Shkarovsky)
$25 from amazon.com **or free online at** makezine.com/go/yakov

When I was a boy my father's best friend, a physicist, gave me a two-volume set called *Physics for Entertainment*. I remember him saying that it was hard to find. All I know is that I loved those books and read them over and over until well past my bedtime.

They explained how everything worked: engines, soap bubbles, eyes, the sun, guns, sailboats, etc., and included neat tricks like how to boil water in a paper cup.

Originally written in 1913 and translated into English in 1975, *Physics for Entertainment* is a fantastic introduction to classical physics. Years ago, I lent my cherished copies to my friend Charles. He returned them to me recently, and flipping through the books for the first time in decades, I was astounded by how many of their images are things that still occur to me frequently.

Charles also told me that *Physics for Entertainment* was republished last year, which is wonderful news. When I looked it up on amazon.com, I learned from comments written by other fans that the books were very popular in India from the 1970s to the 1990s. Interesting!

Hmm, let's see — kids in the U.S.S.R. who read these books started working during the Sputnik era; kids in India who read these books have been joining the country's ever-growing geek force over the past decade — I think I detect a pattern here. And if so, our nation's future just may depend on the new easy availability of Yakov Perelman's *Physics for Entertainment*.

—Paul Spinrad

Decal Master

Monté: King of Atom-Age Monster Decals by Bill Selby
$15 Last Gasp

Every pimply-faced plastic-model enthusiast from the 1950s and 1960s likely remembers "Originals by Monté" decals.

Images on water-slide decals of lurid, slavering monsters, flaming skulls, bleeding eyeballs, and jacked-up street racers were just begging to end up on your models, lunchbox, and bike. It was all the stuff of your mother's nightmares, which only made it cooler.

But who was Monté? Until recently, little was widely known about Don "Monté" Monteverde, now recognized as the crowned king of the water-slide decal. A new book, lovingly produced by Last Gasp (which even includes a Monté decal pack!), is something of an amalgamation of a family scrapbook, art book, and biography.

It tells the rich and ultimately tragic tale of this pioneer of "kustom kulture" art. We learn, among other revelations, that Monteverde was the actual creator of the iconic (and extremely lucrative) Rat Fink character and was hired by Ed "Big Daddy" Roth to do the preliminary sketches (and rarely acknowledged after that).

Eventually, a series of tragedies took the ever-reclusive Monteverde out of the game: the loss of his son in a car accident, the amputation of both of his legs due to poor circulation, and finally, a paralyzing stroke.

Monté may have died in relative obscurity, but he left an indelible mark on underground art and will hopefully now be increasingly acknowledged as the true creator of Rat Fink.

—Gareth Branwyn

PHYSICS FOR ENTERTAIMENT

YAKOV PERELMAN

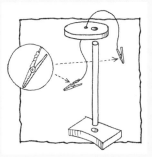

We now have regular tool reviews on the MAKE blog! For your weekly tool fix, check out blog.makezine.com/archive/category/toolbox. Here are a few that recently caught our eye.

1. Pin-Probe Mechanical Stud Finder

$15 garrettwade.com

Garrett Wade's Japanese-made Stud Finder is entirely mechanical, requiring no batteries. It combines a classic magnetic "click" sensor with a spring-loaded pin, shielded by a graduated depth gauge that physically probes the space behind the wall and confirms or denies the presence of a stud pretty much unequivocally. Unlike a capacitive stud finder, it will never have problems working through plaster, foil insulation, or wire lath.

The probe does leave a small hole in the drywall, but it's tiny — a spot of paint will fill it without any spackle at all. Plus, as a bonus feature that's (I think) unique to this type of stud finder, the depth gauge provides an accurate measurement of the thickness of the drywall. This can be a crucial bit of information when choosing fasteners, whether they're screws going into the studs or anchors going into the drywall.

The tool unscrews in the middle, which allows for removal of the pin and/or access to a store of replacement pins in the handle (the Stud Finder ships with 2 pins, and Garrett Wade sells a pack of 10 for $8). The pin plunger has a rotary safety lock to keep the sharp end safely covered when not in use. There's also a lanyard hole in the pommel in case you want to hang it up (which I always do).

—*Sean Michael Ragan*

2. Technician's Pocket Screwdriver

$4 countycomm.com/1x4driver.htm

I have roughly one zillion screwdrivers at my disposal. But the one I use most for servicing small electronics when I've wandered away from my workshop is my Technician's Pocket Screwdriver from County Comm.

It clips to my pants pocket like a pen and is slightly smaller than a Sharpie. I truly appreciate the end caps that protect me from an accidental leg stabbing, and they have side holes so

I can see which tip is which. Each bit is double-ended, giving me a #0 and #00 Phillips choice at one end and a ⅛" and 1⁄16" slotted bit choice at the other. Pull out a bit, flip it around, and push it back into place.

The bits are high-quality hardened steel, but the plastic barrel (and size) means that this isn't a tool for high-torque action. I need it for dealing with tiny little screws that are keeping me out of an enclosure, tuning small pot switches, or tightening down screw terminals on an Arduino project.

When I'm at my workbench, I usually reach for my fancy Wiha screwdrivers, but the one that gets used the most when I'm up and about is this one.

—*John Park*

3. EX210 Mini Digital Multimeter

$70 extech.com

The EX210 multimeter from Extech Instruments is a midrange meter that'll take all the standard hobby readings, such as resistance, continuity, voltage, and current, but also boasts an infrared thermometer and a backlit display.

It came in handy testing the power lines and zipper switch while working on my TV-B-Gone Jacket (blog.craft zine.com).

The website shows off the IR thermometer in a home repair scenario, pointed at a ventilation grate. The built-in laser pointer indicates the target, and the controls are fairly intuitive.

It could also be handy in stovetop candy making, as the thermometer is contact-free, so there's no probe to get all sticky.

Extech's got a great track record for producing quality tools. My only complaint about this model is the size. Even though they call it "mini" (which at Extech distinguishes it from some of the more industrial-level equipment they make), it's still too big to stuff in my bag for a trip to the fabric store (if it's shiny, it might be conductive!).

Every electronics workbench needs a multimeter, and the EX210 is a solid choice.

—*Becky Stern*

Tricks of the Trade By Tim Lillis

Chop that copper.

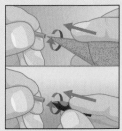

Dave Hrynkiw at Solarbotics shares with us a great trick to cut small-diameter copper tubing without a pipe cutter.

Roll the tubing under a razor blade until it breaks in two. As you roll back and forth, make sure that you're getting a full rotation on the tubing with each stroke.

To clean up the end, you can drag the pipe across coarse sandpaper. Taping the sandpaper down to a table will ensure a stable, level surface.

If you don't have a deburring tool, you can use a cone of rolled up sandpaper or a countersink to smooth the edges.

Have a trick of the trade? Send it to tricks@makezine.com.

Extech Desktop Power Supply
$200 extech.com

Don't let the sexy name fool you. Extech's 382213 Digital Triple Output DC Power Supply is all business — the business of powering your awesome hacks! The 382213 serves up carefully measured voltages to power your electronics projects, reducing the number of batteries you might waste and preventing you from "letting the smoke out" by blasting your components with too much juice.

The 382213 has ports for banana plugs and dials for choosing how much voltage and current you want. It maxes out at 30V and 3A, with LCD displays clearly showing how much of each you've selected. Intriguingly, the spring clips in front dish out fixed voltages, 5V/0.5A and 12V/1A, and quickly catch or release the wires.

I can see the spring clips being great for longer projects where you just want a couple of wires snaking down from your shelf and powering a breadboard without needing to have the power supply cluttering up your desk. *—JB*

John Baichtal is a writer for MAKE, makezine.com, and geekdad.com.

Kindy Connally-Stewart is a 15-year-old sports and science junkie who cut his teeth on Lego and Harry Potter.

Saul Griffith is a columnist of MAKE and chief cyclist at onyacycles.com.

Tim Lillis is a freelance illustrator and DIYer.

Riley Porter is a malware analyst by day and a hacker, maker, and geekdad by night.

Christopher Singleton, a father of three boys, is a maker, inventor, writer, and product development specialist.

Bruce Stewart is a technology writer and contributor to radar.oreilly.com.

Adam Zeloof lives in central New Jersey and enjoys sailing, camping, birding, geocaching, and, of course, making.

Robert M. Zigmund is 13 years old, loves his subscription to MAKE, and has a blog at eztechreviews.blogspot.com.

Zach Zundel is a designer who just started the design firm di|chromate.

✱ Want more? Check out our searchable online database of tips and tools at makezine.com/tnt.

Have a tool worth keeping in your toolbox? Let us know at toolbox@makezine.com.

DANGER!

By Gever Tulley with Julie Spiegler

Squash a Penny on a Train Track

Leverage the force of a locomotive.

1. Pick a location. Find a length of track that is very straight — you want to be able to see and hear the train coming from a long way away. The best location is next to an automated crossing gate — the bells will warn you when a train is coming.

2. Pick a moment. Check the schedule for a gap of at least 15 minutes between trains. Not all train traffic is scheduled, so you must still wait for a time when you can neither see nor hear any trains or crossing bells.

3. Place the penny. Tape the penny to the top of the rail, to prevent the vibration of the approaching train from shaking the penny off the rail. If there's a bright, shiny part of the rail, tape the penny there; that's where the train makes the best contact with the rail.

4. Stand back and wait. Stand at least 30 feet away from *all* tracks, and wait for a train to pass. If the tape doesn't hold the penny in place, it may come flying out at high velocity.

5. Find the penny. After the train passes, and you can neither see nor hear any trains or crossing bells, find the squashed penny. Be careful, it may still be hot from being squashed. Get away from the track as soon as you have your penny.

Using the tape, it may be possible to get two different types of coins to squash together. Preparing them by sanding their surfaces will increase the chances of a good metallurgical bond. To ensure that you don't harm the train or the track, *never* put anything larger than a coin or two on the tracks. ▪

GET A TICKET

PROJECTILES

RUN OVER BY TRAIN

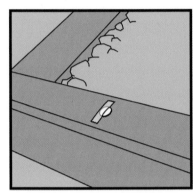

⚠ **WARNING:** Because of the unfamiliar size of train engines, our brains don't accurately judge their distance and speed. If you see or hear a train, assume that it's a danger and move to a safe distance immediately. Also, coins may squirt out from under train wheels, so stand at least 30 feet away from where you placed your penny.

REQUIRES	DURATION
Pennies or other coins	**Short**
Tape	
Active train track	DIFFICULTY
Train schedule	**Easy**

SUPPLEMENTARY DATA

Trains have no steering wheel.

The faster you want a train to go, the smoother the track has to be. If the pieces of the track are bolted together with plates, then only slow trains use it; if they're welded into a continuous piece of rail, the track is probably designed for faster trains.

While a coin has never derailed a train, more than one person has been injured putting coins on a rail. Usually, it's because they accidentally stand on another track while waiting, and a train comes down that other track and hits them.

Excerpted from *Fifty Dangerous Things (You Should Let Your Children Do)* by Gever Tulley with Julie Spiegler (fifty dangerousthings.com). Gever is co-founder of Brightworks, a new K–12 school in San Francisco (sfbrightworks.org).

Gever Tulley

→ While working on this issue, we hosted a simple little contest, putting out an all-points bulletin in search of robots with lots of character and personality. Contestants were asked to submit their build photos and instructions on our DIY wiki, Make: Projects (makeprojects.com). And since robot personalities come through best in action, submitting a video of the bot was mandatory. Because our community rocks, we got some great entries. Thanks to everyone who entered! Deciding the winners was not easy, but here they are (drum roll please):

TOP BOT

"Chopsticks" the Spider Robot
By Russell Cameron
makeprojects.com/Project/748/1

This little bot has tons of character chops, and we love his legs made of Polymorph plastic and chopsticks. Check out the vid of him contemplating his own end effectors and playing peekaboo.

RUNNERS-UP

RoboBrrd
By Erin Kennedy
makeprojects.com/Project/862/1

Bio-inspired design combined with sensing, interesting behaviors, and a colorful design caught our eye. Erin's documentation is also top-notch. Bonus points for the eraser eyelashes.

Robot Drummer (of Double Rainbow Band)
By Tim Laursen
makeprojects.com/Project/800/1

We're not sure we've ever seen a bot look like this before (or incorporate this many colors), so the Drummer is certainly original. While not super complex, he's definitely entertaining.

Belvedere, a Butler Robot
By Andy Wolff
makeprojects.com/Project/823/1

Who doesn't need a butler bot? Belvedere can navigate the rooms of his house, tell jokes, serve food and drinks, play music, and dance. No shortage of character here!

what I made

ELEGANT TELESCOPIC LIGHT WITH WEIGHTED COUNTERBALANCE

MADE ENTIRELY FROM PAPER AND CARD (AND A TEENY BIT OF WIRE AND OLD XMAS LIGHTS) THIS COOL SHADE NEEDS NOTHING MORE THAN A CRAFT KNIFE, A HOT GLUE GUN, AND A PAIR OF PLIERS TO CONSTRUCT.

5mm
FOAMCORE BOARD

A STRING OF 80 LED LIGHTS

1.5mm WIRE

DOUBLE DIFFUSED WHEN DOWN.

SINGLE DIFFUSED WHEN UP.

① START HERE

FIND 2 OLD 'WRAPPING PAPER' TUBES WITH DIFFERENT DIAMETERS - ONE NEEDS TO FIT INSIDE THE OTHER. CUT THEM TO THE LENGTHS SHOWN. NOW CUT 2 DISCS OF FOAMCORE BOARD 180mm IN DIAMETER. CUT A HOLE IN ONE 5mm WIDER THAN THE NARROWEST TUBE.

180mm

OUTSIDE TUBE - 590mm

INSIDE TUBE - 575mm

PAPER AND CARD GLUE

④

IF NECESSARY, GLUE STRIPS OF CARD AROUND THE TOP OF THE INNER TUBE UNTIL IT FITS SNUGLY WITHIN THE OUTER TUBE. NOW USE 1.5mm GALVANISED WIRE TO MAKE THE HANDLE AND GUIDE WIRE AS SHOWN BELOW. ATTACH THEM TO THE TUBES USING HOT GLUE...

TOP VIEW OF GUIDE WIRE

GUIDE WIRE

HOT GLUE

HANDLE

HANDLE

SPACER CARD

90°

②

GLUE THE 2 DISCS TOGETHER USING PAPER AND CARD GLUE. PLACE THE NARROW TUBE IN THE CENTRE OF THE HOLE IN THE TOP DISC. FILL THE GAP WITH HOT GLUE, ENSURING THE TUBE IS VERTICAL (USE A SPIRIT LEVEL).

70 ± mm

25 mm

4 STRANDS OF SEWING THREAD.

③

GLUE COINS TOGETHER TO CREATE A WEIGHT. CREATE A HOOK FROM A PAPER CUP AND HOT-GLUE IT ONTO ONE END. ATTACH 70CM OF SEWING THREAD (USE 4 STRANDS) TO THE HOOK USING A KNOT.

⑤ ... BUT LOWER THE WEIGHT INTO THE INNER TUBE FIRST! THEN, WITH THE THREAD RUNNING OVER THE GUIDE WIRE AND DOWN THE OUTSIDE OF THE INNER TUBE (AND WITH THE WEIGHT AT THE TOP) LOWER THE OUTER TUBE, KEEPING THE THREAD TAUT, UNTIL IT RESTS ON THE TOP OF THE BASE. USE HOT GLUE TO ATTACH THREAD TO THE BOTTOM OF THE OUTER TUBE.
BE CAREFUL NOT TO TWIST THE TUBES, OTHERWISE THE THREAD WILL JAM OR SNAP.

SEWING THREAD

GLUED

SLACK

⑥

TAKE THE LED LIGHTS AND WIND THEM AROUND THE OUTER TUBE. USE HOT GLUE TO HOLD THEM IN PLACE. BEND EACH LED LIGHT OUTWARD. IF THERE'S A 'SEQUENCER' BOX GLUE IT ONTO THE BASE AND MAKE SURE THERE'S ENOUGH...

... SLACK WIRE FOR THE OUTER TUBE TO MOVE FREELY UP AND DOWN.

GLUE

⑦

TAKE SOME A2 150g CARD AND CUT IT INTO A PIECE 590mm x 575mm. CRUMPLE IT AND THEN SMOOTH IT OUT. WRAP IT AROUND THE BASE AND GLUE INTO POSITION. FOR ADDITIONAL SUPPORT, BEND WIRE INTO A

WIRE RING

GLUE

RING AND GLUE ONTO THE INSIDE EDGE OF CARD.

⑧

CUT A FOAMCORE DISC 250 mm IN DIAMETER. CUT A HOLE IN THE CENTRE WIDER THAN THE OUTER TUBE. WRAP WITH CRUMPLED PAPER 590mm x 654mm IN SIZE (SEE PREVIOUS STEP). GLUE A CIRCULAR DISC, WITH A SLIT FOR THE HANDLE, OVER THE HOLE. TO FINISH OFF, GLUE A WIRE RING ONTO THE INSIDE EDGE OF THE CARD.

FINALLY, MAKE 4 FEET FROM BENT WIRE AND HOT-GLUE THEM ONTO THE BASE. GOOD LUCK!

BY SCOTT BEDFORD
WHATIMADE.COM

WINNER
THE 15TH ANNUAL WEBBY AWARDS

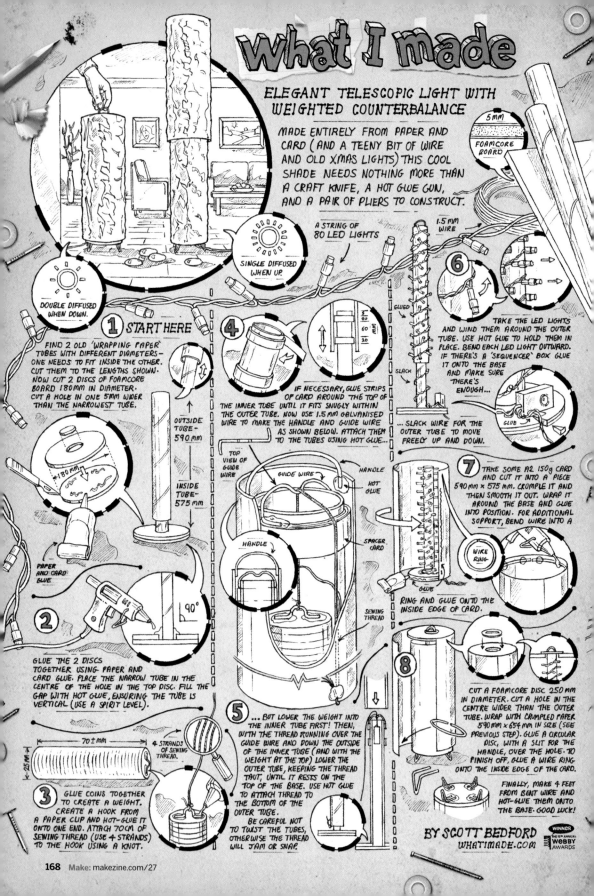

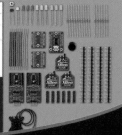

The Daniell Cell

Make the electric battery that powered the scientific revolution.

SEVERAL OF THE PROJECTS FEATURED previously in this column were battery powered. Humphrey Davy, who developed the arc light (Volume 20), and Samuel Morse, inventor of the telegraph (Volume 23), used voltage produced by chemical reaction.

In 1800 Alessandro Volta made the first apparatus that turned chemical energy into a stable, long-lasting, and constant electrical voltage (Volume 24). But it was hard to use because the chemical reaction quickly corroded the parts.

In 1836, English chemist John Frederic Daniell invented a dependable, easy-to-use battery, so good that it was used in one shape or another until the 1950s.

Like Benjamin Franklin, Daniell was a polymath and prolific inventor, as well as a horticulturalist and meteorologist. He taught chemistry, invented heat and humidity gauges, and wrote the *Introduction to Chemical Philosophy* (1839).

Inside the Daniell cell, electrons are transferred from the zinc electrode to the copper one, producing slightly more than 1 volt. Unlike Volta's early batteries, the Daniell cell made use of 2 separate but electrically connected electrolyte solutions. This method radically lengthened the life of the cell.

In this edition of Remaking History, we build a Daniell cell battery using common hardware-store chemicals.

No matter how big or small it's made, each zinc-copper Daniell cell produces about 1.1 volts. It takes 2 Daniell cells connected in series to power an LED.

JUICE MAKER

✎ English polymath John Frederic Daniell (1790–1845) invented the Daniell electric cell in 1836, a radical improvement on the Volta cell. His brilliant career ended when he died of apoplexy at 55 at a meeting of the Royal Society, to which he'd been elected at 23.

MATERIALS

For each Daniell cell:

Copper strip, ¾"×3" A 3"-long copper tube will also work.

Zinc strip, ¾"×3" You can buy zinc sheet metal at some hardware stores. Alternatively, you can harvest zinc from the outer shell of a D-size non-alkaline battery. Carefully open the battery with a hacksaw or rotary tool (Dremel) equipped with an abrasive cutting-wheel attachment. Scoop out the black powder with a spoon and remove the carbon rod inside. Clean the zinc and then cut it to size.

Mason jars and lids, quart-size (2)

Plastic hose, 1" diameter, 8" long

Cotton balls (6)

Copper sulfate In its pentahydrate form, it's typically found in the drain cleaner section of hardware stores. Manufacturers include Rooto and Roembic.

Zinc sulfate Available in garden centers and hardware stores as moss remover.

Table salt and distilled water

Alligator clip leads (2)

LED (optional)

TOOLS

Voltmeter

Glass stirring rod

Scale for weighing chemicals

Getty

→ START

1. Make the copper sulfate solution.
In a Mason jar, dissolve about 50 grams of copper sulfate in a pint of distilled water. The copper sulfate will be slow to dissolve, but if you stir or shake the covered jar long enough, most of it will eventually go into solution.

2. Make the zinc sulfate solution.
In a second Mason jar, dissolve 50 grams of zinc sulfate in a pint of distilled water.

3. Make the salt water solution.
In a bowl, dissolve 25 grams of table salt in a cup of distilled water.

4. Make the saline bridge.
Place two cotton balls in one end of the 1"-diameter plastic tube. Fill the tube completely with salt water, and plug the other end with 2 more cotton balls.

5. Position the saline bridge.
Put one end of the bridge in each of the Mason jars. Note that some salt water may leak out of the bridge. Some leakage is acceptable; however, there must be a continuous salt water connection from one Mason jar to the other.

6. Insert the metal strips.
Attach the metal strips to the alligator clips and voltmeter. Place the copper metal strip in the copper sulfate solution and the zinc strip in the zinc solution (Figure A). With the bridge in place, the voltmeter should read 1.0 to 1.1 volts (Figure B). You've got a Daniell cell.

7. Prepare a second Daniell cell.

8. Attach the 2 Daniell cells in series.
Connect cell 1's zinc strip to cell 2's copper strip. Then connect cell 1's copper to the LED's positive (long) lead, and cell 2's zinc to its negative (short) lead (Figure C). If you've done it right, the LED will light. Congratulations, you've made a battery! ☒

William Gurstelle is a contributing editor of MAKE. His new book, *The Practical Pyromaniac*, is available in the Maker Shed (makershed.com) and at other fine booksellers.

Gregory Hayes; Gerry Arrington (C)

Unlike Volta's cell, Daniell's used 2 separate electrolyte solutions, which radically lengthened its life.

⚠ **CAUTION:** Wear rubber gloves and splash-proof eye protection when mixing chemical solutions. Do NOT pour any chemical solutions down the storm sewer. Empty them in a utility drain or toilet.

A

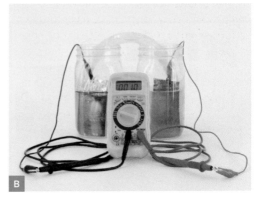

B

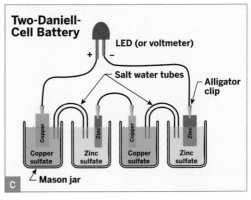

Two-Daniell-Cell Battery

LED (or voltmeter)

+ −

Salt water tubes

Alligator clip

Copper — Zinc — Copper — Zinc

Copper sulfate — Zinc sulfate — Copper sulfate — Zinc sulfate

Mason jar

C

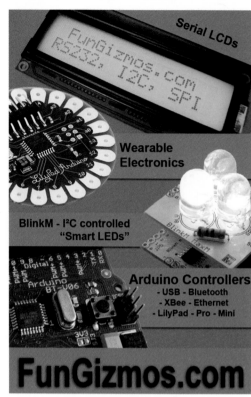

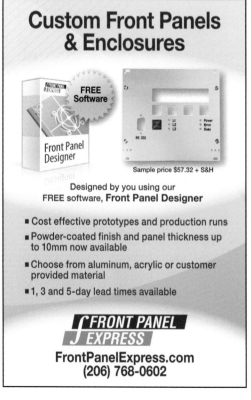

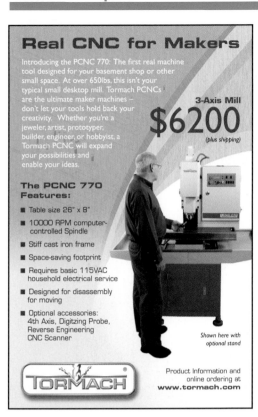
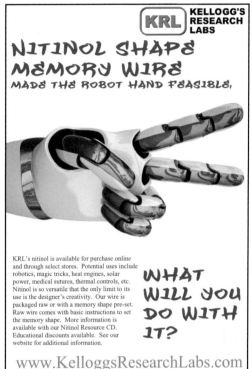

HOMEBREW

My DIY King Pong
By Jerry Reilly

SITTING AROUND A CAMPFIRE LAST summer, we decided the world really needed a gigantic outdoor version of *Pong* that could turn up in the most unlikely places. Thus, King Pong was born. (For those born after 1972, *Pong* was the first popular video game — just two moving paddles and a bouncing ball.)

The pieces we needed to put together were a netbook computer running a custom version of *Pong*, a video projector, a portable power supply, and two game controllers. The first job was to write a clone of the original *Pong* game in Visual Basic. For the projector, we searched for the most lumens per dollar and chose a Dell 1510X. To power it all, we got an old marine battery and a power inverter.

For the controllers, we wanted something that would grab attention and be easily recognizable. Ideally it would be wireless, as we learned from our tests that wires in the dark were a constant menace. So we mounted a bicycle wheel on a cheapo camera tripod using a standard L-bracket.

The sensor that detects the wheel's motion is a simple single-pole double-throw (SPDT) switch mounted so that it sticks out between the spokes. As you spin the wheel in either direction, one of the pairs of contacts toggles on/off. Finally, we used a pair of XBee wireless modules, one for each controller, to send the signal back to the netbook (aka the Pongputer).

With this setup, we have a portable system that can be thrown in the back seat of a car, set up anywhere in less than five minutes, and project a 100-foot-high *Pong* screen on nearly any surface. King Pong has filled an enormous sand dune on Cape Cod; it's been projected from the top of a bridge across a river onto trees in a forest; it's been played on the surface of a pond. You never know where King Pong might turn up next. ◪

Jerry Reilly entertains himself and the public with the various activities of Pedestrian Magic. For VB source code, email jerry@pedestrianmagic.org.

Jerry Reilly